THE MOON LANDING HOAX

THE MOON LANDING HOAX
The Eagle That Never Landed

DR STEVEN THOMAS

(c) 2010 DR STEVEN THOMAS

All rights reserved

No part of this publication may be produced, stored in a retrieval system, or transmitted in any form or by any means, without the prior permission in writing of the publisher, nor be otherwise circulated in any form of binding or cover other than in which it is published and without a similar condition including this condition being imposed on the subsequent purchaser.

First published in Great Britain by Swordworks Books

ISBN 978-1-906512-47-7

Printed and bound in the UK & US

A catalogue record of this book is available from the British Library

Cover design by Swordworks Books

CONTENTS

PART 1	7
PART 2	37
PART 3	77
GLOSSARY	91

PART 1

THE APOLLO PROGRAM

"One small step for man, one giant leap for mankind." Neil Armstrong July 20, 1969

The sixties and seventies saw the cold war reaching new heights, and the supremacy over space was too important for both United states as well as the Soviet republic. President John F. Kennedy had made a public commitment on 25 May 1961 to land an American on the Moon by the end of the decade, but up until this time, Apollo had been all promise. Kennedy's decision had involved much study and review prior to making it public, and his commitment had captured the American imagination, generating overwhelming support. Project Apollo had originated as an effort to deal with the unsatisfactory situation i.e. the world perception of Soviet leadership in space and technology, and it addressed these problems very well. Even though Kennedy's political objectives were essentially achieved with the decision to go to the Moon, Project Apollo took on a life of its own over the years and left an important legacy to both the nation and the proponents of space exploration. Its success was enormously significant, coming at a time when American society was in crisis.

NASA had the responsibility to accomplish the task set out in a few short paragraphs by the president.

NASA in the 1960s, image courtesy NASA

PART 1

The Apollo program was the result of many years of effort and expense. There was tremendous pressure on all those involved as the race began to put the first man on the moon. The first Apollo program was delayed because of a tragedy due to a devastating fire inside the capsule. All astronauts died in this tragic accident on the launch pad. Among the various missions Apollo 6 and Apollo 8 have more prominence. The Apollo 6 space mission was launched from the Kennedy Space Center, Florida. The liftoff on April 4, 1968, was the last unmanned test of the huge Saturn V rocket systems and the Apollo modules. The other Apollo flights up to this time tested different parts of the complex project. The crew of Apollo 8 were the first humans to orbit the moon, They orbited it 10 times in December, 1968.

On July 16, 1969, the Apollo 11 launched from the Kennedy Space Center. The crew consisting of Neil Armstrong, Commander; Michael Collins, command module pilot; and , Edwin E. Aldrin Jr., lunar module pilot set off to land on the moon. On July 20, 1969, Commander Neil Armstrong succeeded in his mission. A camera in the Lunar Module provided live television coverage as Neil Armstrong climbed down the ladder to the surface of the moon.

The Lunar Module "Eagle" consisted of two parts: the descent stage and the ascent stage.

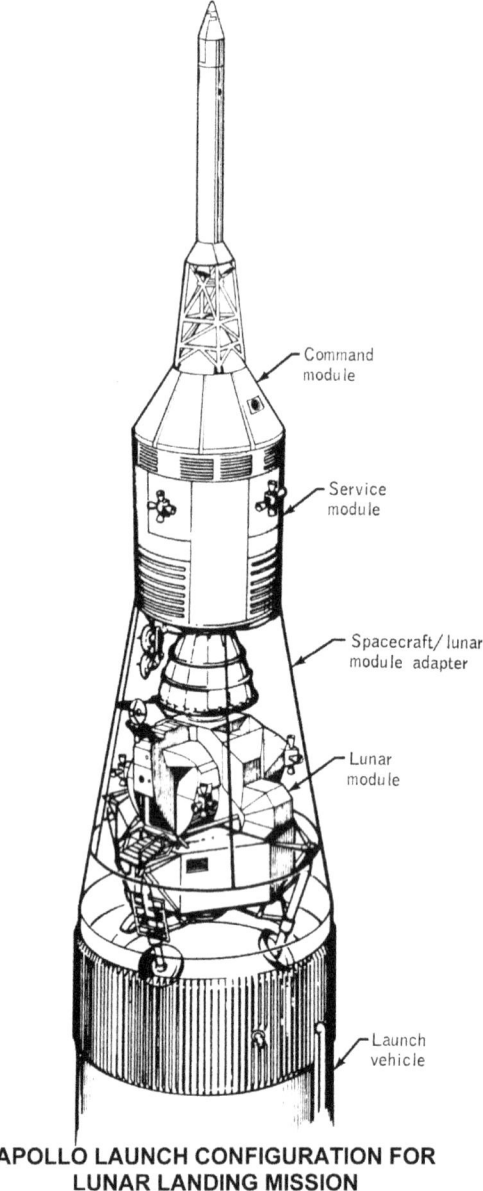

**APOLLO LAUNCH CONFIGURATION FOR
LUNAR LANDING MISSION**

PART 1

The descent state provided the engine used to land on the moon. It had four legs, a storage area for experimental gear, and a ladder for the crew to climb down to the moon's surface. The descent module also served as the launch platform for the ascent module when it came time to leave. The ascent module carried the crew back to the Command Service Module.

To walk on the moon's surface, the astronauts needed to wear a space suit with a back mounted, portable life support system. This controlled the oxygen, temperature and pressure inside the suit.

Space suit designs, image courtesy NASA

THE MOON LANDING HOAX

On the surface, the astronauts had to get used to the reduced gravity. They could jump very high compared to on Earth. The crews spend a total of two and a half hours on the moon's surface. While on the moon's surface, they performed a variety of experiments and collected soil and rock samples to return to Earth.

Flag on the moon, image courtesy NASA

An American flag was left on the moon's surface as a reminder of the accomplishment.

After accomplishing their task they set back to the earth and on re-entering the Earth's atmosphere, parachutes opened to safely lower the Columbia into the Pacific Ocean. After landing in the Ocean, the crew were

PART 1

retrieved by a helicopter and taken to the recovery ship, the "USS Hornet." The crew and lunar samples were placed in quarantine until their health and safety could be confirmed.

The Command Module "Columbia" returned to Earth on July 24, 1969. President Kennedy's objective to land men on the moon and return them safely to Earth had been accomplished.

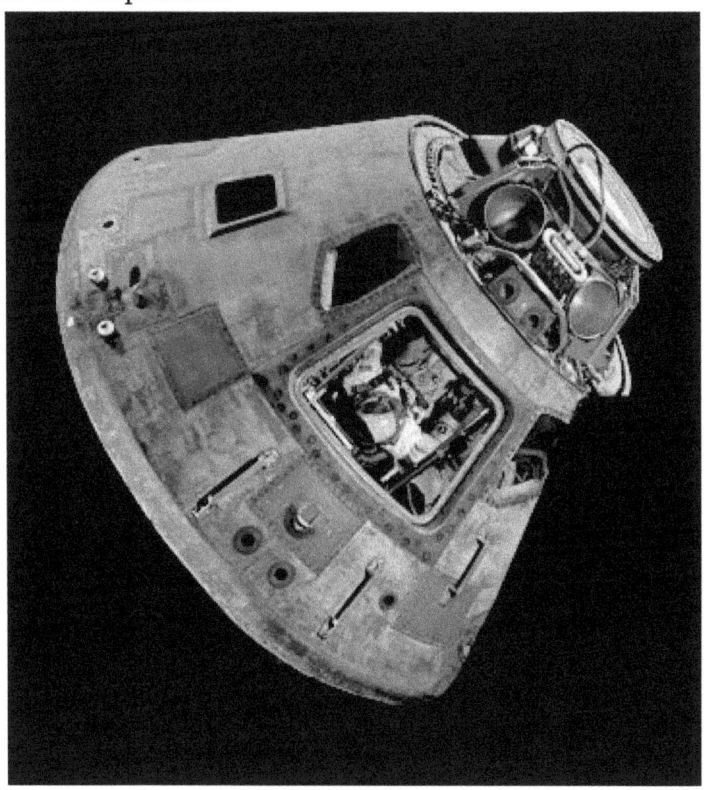

Columbia Command Module, image courtesy NASA

An ecstatic reaction enveloped the globe, as everyone shared in the success of the mission. Ticker tape parades, speaking engagements, public relations events, and a world tour by the astronauts served to create good will both in the United States and abroad.

Five more landing missions followed at approximately six-month intervals through December 1972, each of them increasing the time spent on the Moon. The scientific experiments placed on the Moon and the lunar soil samples returned have provided grist for scientists' investigations ever since.

Three of the later Apollo missions used a lunar rover vehicle to travel in the vicinity of the landing site, but none of them equaled the excitement of Apollo 11.

Project Apollo in general and the flight of Apollo 11 in particular, should be viewed as a watershed in American history. It was an endeavor that demonstrated both the technological and economic virtuosity of the United States and established national preeminence over rival nations—the primary goal of the program when first envisioned by the Kennedy administration in 1961.

It had been an enormous undertaking, costing $25.4 billion i.e about $95 billion in 1990 dollars, with only the building of the Panama Canal rivaling the Apollo program's size as the largest non-military technological endeavor ever undertaken by the United States and only the Manhattan Project being comparable in a wartime.

PART 1

Lunar Rover, image courtesy NASA

First, and probably most important, the Apollo program was successful in accomplishing the political goals for which it had been created. Kennedy had been dealing with a Cold War crisis in 1961 brought on by several separate factors—the Soviet orbiting of Yuri Gagarin and the disastrous Bay of Pigs invasion only two of them—that Apollo was designed to combat. At the time of the Apollo 11 landing, Mission Control in Houston flashed the words of President Kennedy announcing the Apollo commitment on its big screen. Those phrases were followed with these: "TASK ACCOMPLISHED, July 1969." No greater understatement could probably have been made. Any assessment of Apollo that does not recognize the accomplishment of landing an American on the Moon and safely returning before the end of the 1960s is incomplete and inaccurate, for that was the primary goal of the undertaking.

HISTORY OF NASA

> *"An Act to provide for research into the problems of flight within and outside the Earth's atmosphere, and for other purposes."*

With this simple preamble, the Congress and the President of the United States created the national Aeronautics and Space Administration in short NASA on October 1, 1958. NASA's birth was directly related to the pressures of national defense. After World War II, the United States and the Soviet Union were engaged in the Cold War, a broad contest over the ideologies and allegiances of the nonaligned nations. During this period, space exploration emerged as a major area of contest and became known as the space race.

PART 1

During the late 1940s, the Department of Defense pursued research and rocketry and upper atmospheric sciences as a means of assuring American leadership in technology. A major step forward came when President Dwight D. Eisenhower approved a plan to orbit a scientific satellite as part of the International Geophysical Year for the period, July 1, 1957 to December 31, 1958, a cooperative effort to gather scientific data about the Earth. The Soviet Union quickly followed suit, announcing plans to orbit its own satellite.

The Naval Research Laboratory's Project Vanguard was chosen on 9 September 1955 to support the IGY effort, largely because it did not interfere with high-priority ballistic missile development programs. It used the non-military Viking rocket as its basis while an Army proposal to use the Redstone ballistic missile as the launch vehicle waited in the wings. Project Vanguard enjoyed exceptional publicity throughout the second half of 1955, and all of 1956, but the technological demands upon the program were too great and the funding levels too small to ensure success.

A full-scale crisis resulted on October 4, 1957 when the Soviets launched Sputnik 1, the world's first artificial satellite as its IGY entry. This had a "Pearl Harbor" effect on American public opinion, creating an illusion of a technological gap and provided the impetus for increased spending for aerospace endeavors, technical and scientific educational programs, and the chartering of new federal agencies to manage air and space research and development.

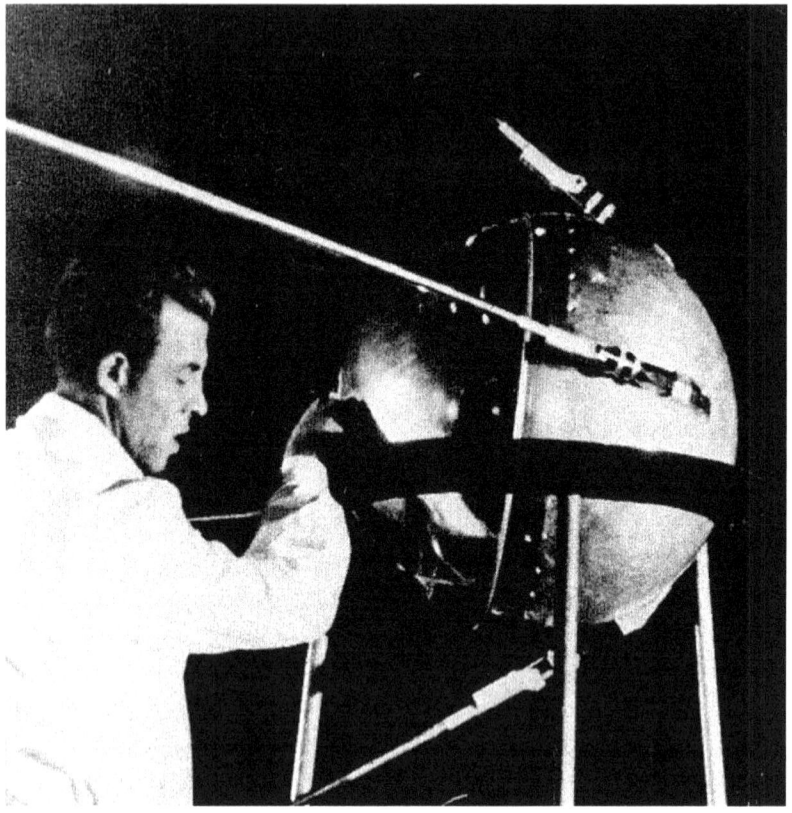

Sputnik

More immediately, the United States launched its first Earth satellite on January 31, 1958, when Explorer 1 documented the existence of radiation zones encircling the Earth. Shaped by the Earth's magnetic field, what came to be called the Van Allen Radiation Belt, these zones partially dictate the electrical charges in the atmosphere and the solar radiation that reaches Earth. The U.S. also began a series of scientific missions to the Moon and planets in the latter 1950s and early 1960s.

A direct result of the Sputnik crisis, NASA began operations on October 1, 1958, absorbing into itself the earlier National Advisory Committee for Aeronautics intact: its 8,000 employees, an annual budget of $100 million, three major research laboratories-Langley Aeronautical Laboratory, Ames Aeronautical Laboratory, and Lewis Flight Propulsion Laboratory-and two smaller test facilities. It quickly incorporated other organizations into the new agency, notably the space science group of the Naval Research Laboratory in Maryland, the Jet Propulsion Laboratory managed by the California Institute of Technology for the Army, and the Army Ballistic Missile Agency in Huntsville, Alabama, where Wernher von Braun's team of engineers were engaged in the development of large rockets. Eventually NASA created other Centers and today it has ten located around the country.

NASA began to conduct space missions within months of its creation, and during its first twenty years NASA conducted several major programs

NASA's first high-profile program involving human spaceflight was Project Mercury, an effort to learn if humans could survive the rigors of spaceflight. On May 5, 1961, Alan B. Shepard Jr. became the first American to fly into space, when he rode his Mercury capsule on a 15-minute suborbital mission. John H. Glenn Jr. became the first U.S. astronaut to orbit the Earth on February 20, 1962. With six flights, Project Mercury achieved its goal of putting piloted spacecraft into Earth orbit and retrieving the astronauts safely.

Project Gemini, image courtesy NASA

PART 1

Project Gemini built on Mercury's achievements and extended NASA's human spaceflight program to spacecraft built for two astronauts. Gemini's 10 flights also provided NASA scientists and engineers with more data on weightlessness, perfected reentry and splashdown procedures, and demonstrated rendezvous and docking in space. One of the highlights of the program occurred during Gemini 4, on June 3, 1965, when Edward H. White, Jr., became the first U.S. astronaut to conduct a spacewalk.

Edward White, first spacewalk, image courtesy NASA

The singular achievement of NASA during its early years involved the human exploration of the Moon, Project Apollo. Apollo became a NASA priority on May 25 1961, when President John F. Kennedy announced "I believe that this nation should commit itself to achieving the goal, before this decade is out, of landing a man on the Moon and returning him safely to Earth." A direct response to Soviet successes in space, Kennedy used Apollo as a high-profile effort for the U.S. to demonstrate to the world its scientific and technological superiority over its cold war adversary. In response to the Kennedy decision, NASA was consumed with carrying out Project Apollo and spent the next 11 years doing so. On July 20, 1969, the Apollo 11 mission fulfilled Kennedy's challenge by successfully landing Armstrong and Edwin E. "Buzz" Aldrin, Jr. on the Moon.

Five more successful lunar landing missions followed. The Apollo 13 mission of April 1970 attracted the public's attention when astronauts and ground crews had to improvise to end the mission safely after an oxygen tank burst midway through the journey to the Moon. Although this mission never landed on the Moon, it reinforced the notion that NASA had a remarkable ability to adapt to the unforeseen technical difficulties inherent in human spaceflight.

With the Apollo 17 mission of December 1972, NASA completed a successful engineering and scientific

PART 1

program. Fittingly, Harrison H. "Jack" Schmitt, a geologist who participated on this mission, was the first scientist to be selected as an astronaut. NASA learned a good deal about the origins of the Moon, as well as how to support humans in outer space. In total, 12 astronauts walked on the Moon during 6 Apollo lunar landing missions.

Apollo 17, the final mission, image courtesy NASA

THE APOLLO 11 ASTRONAUTS

The Astronauts for Apollo 11 consisted of Edwin E. Aldrin, Jr., Neil Armstrong and Michael Collins.

Edwin E. Aldrin, Jr. was born in Montclair, New Jersey, on 20 January 1930. He attended the U.S. Military Academy at West Point, entered the United States Air Force, and received pilot training in 1951.

Buzz Aldrin, image courtesy NASA

PART 1

Aldrin flew sixty-six combat missions in F-86s in Korea, destroying two MIG-15 aircraft. Known to all as by his nickname, "Buzz," Aldrin was also one of the most important figures in the accomplishment of Project Apollo in successfully landing an American on the Moon in 1960s.

Aldrin became an astronaut during the selection of the third group by NASA in October 1963. On 11 November 1966 he orbited aboard the Gemini XII spacecraft, a 4-day 59-revolution flight that successfully ended the Gemini program. It proved to be a fortuitous selection, for during Project Gemini Aldrin became one of the key figures working on the problem of rendezvous of spacecraft in Earth or lunar orbit, and docking them together for spaceflight. Without these skills Apollo could not have been successfully completed.

Aldrin, with a Ph.D. in astronautics from Massachusetts Institute of Technology, was ideally qualified for this work, and his intellectual inclinations ensured that he carried out these tasks with enthusiasm. Systematically and laboriously, Aldrin worked to develop procedures and tools necessary to accomplish space rendezvous and docking. He was also a central figure in devising the methods necessary to carry out extravehicular activities of astronauts outside their vehicles. That, too, was critical to the successful accomplishment of Apollo. Aldrin was the second American to set foot on the lunar surface. He and Apollo 11 commander Neil A. Armstrong spent about two and half hours on the Moon before returning to the orbiting Apollo Command Module. The spacecraft and the lunar explorers returned to Earth on 24 July 1969.

In 1971 Aldrin returned to the Air Force and retired a year later. He wrote two important books about his activities in the U.S. space program. In Return to Earth, Aldrin recounted the flight of Apollo 11. In the more broadly constructed Men from Earth , Aldrin discussed the entire space race between the United States and the Soviet Union. He has been an important analyst of the space program since the 1960s.

Neil Alden Armstrong was born on 5 August 1930 on his grandparents' farm near Wapakoneta, Ohio, to Stephen and Viola Armstrong. Because Armstrong's father was an auditor for the State of Ohio, Armstrong grew up in several communities, including Warren, Jefferson, Ravenna, St. Marys, and Upper Sandusky, before the family settled in Wapakoneta.

Armstrong developed an interest in flying at only age two when his father took him to the National Air Races in Cleveland, Ohio. His interest intensified when he went for his first airplane ride in a Ford Tri-Motor, a "Tin Goose," in Warren, Ohio, at age six. From that time on, he claimed an intense fascination with aviation.

PART 1

Neil Armstrong, image courtesy NASA

At age fifteen, Armstrong began taking flying lessons at an airport north of Wapakoneta, working at various jobs in town and at the airport to earn the money for lessons in an Aeronca Champion airplane. By age sixteen, he had his student pilot's license, before he even passed his automobile driver's test and received that license and before he graduated from Blume High School in Wapakoneta in 1947.

Immediately after high school Armstrong received a scholarship from the U.S. Navy. He enrolled at Purdue University and began his studies of aeronautical engineering. In 1949, the Navy called him to active duty, where he became an aviator, and in 1950, he was sent to Korea. There he flew seventy-eight combat missions from the aircraft carrier U.S.S. Essex.

After mustering out of the Navy in 1952, Armstrong joined the National Advisory Committee for Aeronautics. His first assignment was at the NACA's Lewis Research Center, near Cleveland, Ohio. For the next seventeen years, he was an engineer, test pilot, astronaut, and administrator for the NACA and its successor agency, the National Aeronautics and Space Administration.

In the mid-1950s Armstrong transferred to NASA's Flight Research Center, Edwards, California, where he became a research pilot NACA's High-Speed Flight Station at Edwards Air Force Base in California as an aeronautical research scientist and then as a pilot on many pioneering high-speed aircraft, including the well-known, 4,000-mph X-15. He flew over 200 different models of aircraft, including jets, rockets, helicopters, and gliders. While there he also pursued graduate studies, and received a master of science degree in aerospace engineering from the University of Southern California.

Armstrong transferred to astronaut status in 1962, one of nine NASA astronauts in the second class to be chosen. He moved to El Lago, Texas, near Houston's Manned Spacecraft Center, to begin his astronaut training. There he underwent four years of intensive training for the Apollo program to land an American on the Moon before the end of the decade.

On 16 March 1966, Armstrong flew his first space mission as command pilot of Gemini VIII with David Scott. During that mission Armstrong piloted the Gemini VIII spacecraft to a successful docking with an Agena target spacecraft already in orbit. While the docking went smoothly and the two craft orbited together, they began to pitch and roll wildly. Armstrong was able to undock the Gemini and used the retro rockets to regain control of his craft, but the astronauts had to make an emergency landing in the Pacific Ocean.

As spacecraft commander for Apollo 11, the first piloted lunar landing mission, Armstrong gained the distinction of being the first person to step on the surface of the

Moon. On 16 July 1969, Armstrong, Michael Collins, and Edwin E. "Buzz" Aldrin began their trip to the Moon. Collins was the Command Module pilot and navigator for the mission. Aldrin, a systems expert. As commander of Apollo 11, Armstrong piloted the Lunar Module to a safe landing on the Moon's surface. On 20 July 1969, at 10:56 p.m. EDT, Neil Armstrong stepped down onto the Moon and made his famous statement, "That's one small step for a man, one giant leap for mankind." Armstrong and Aldrin spent about two and one-half hours walking on the Moon collecting samples, doing experiments, and taking photographs. On 24 July 1969, the three men splashed down in the Pacific Ocean. They were picked up by the aircraft carrier, U.S.S. Hornet.

The three Apollo 11 astronauts were honored with a ticker tape parade in New York City soon after returning to Earth. Armstrong received the Medal of Freedom, the highest award offered to a U.S. civilian. Armstrong's other awards coming in the wake of the Apollo 11 mission included the NASA Distinguished Service Medal, the NASA Exceptional Service Medal, seventeen medals from other countries, and the Congressional Space Medal of Honor.

Armstrong subsequently held the position of Deputy Associate Administrator for Aeronautics, NASA Headquarters, Washington, D.C., in the early 1970s. In that position, he was responsible for the coordination and management of overall NASA research and technology work related to aeronautics.

After resigning from NASA in 1971, he became a professor of Aerospace Engineering at the University of Cincinnati from 1971 to 1979. During the years 1982-1992, Armstrong served as chairman of Computing Technologies for Aviation, Inc., in Charlottesville, Virginia. He then became chairman of the board of AIL Systems, Inc., an electronics systems company in Deer Park, New York.

Michael Collins was born on October 30, 1930, in Rome, Italy. He later moved to Washington, D.C., where he graduated from St. Albans School. In 1952, he attended the U.S. Military Academy at West Point, New York, and received his bachelor of science degree.

Prior to joining NASA, Collins served as a fighter pilot and an experimental test pilot at the Air Force Flight Center, Edwards Air Force Base, California. From 1959 to1963 he logged more than 4,200 hours of flying time.

In October 1963, Michael Collins became one of the third group of astronauts named by NASA. He served as a pilot on the three-day Gemini X mission, launched July 18, 1966. During this mission, he set a world altitude record and became the nation's third spacewalker while completing two extravehicular activities.

His second flight was as Command Module pilot of the historic Apollo 11 mission in July 1969. He remained

THE MOON LANDING HOAX

in lunar orbit while Neil Armstrong and Buzz Aldrin became the first people to walk on the Moon. His role in the Apollo mission earned him many awards and accolades, including the Presidential Medal for Freedom in 1969.

Michael Collins, image courtesy NASA

PART 1

In January 1970, Collins left NASA to become the Assistant Secretary of State for Public Affairs. A year later he joined the Smithsonian Institution as the Director of the National Air and Space Museum, where he remained for seven years. While in this position, he was responsible for the construction of the new museum building, which opened to the public in July 1976, ahead of schedule and below its budgeted cost. In April 1978, Collins became Under Secretary of the Smithsonian Institution.

In 1980, he became the Vice President of the LTV Aerospace and Defense Company, resigning in 1985 to start his own firm.

Collins has completed two spaceflights, logging 266 hours in space, of which 1 hour and 27 minutes was spent in EVA. He has written about his experiences in the space program in several books, including Carrying the Fire and Flying to the Moon and other Strange Places. In 1988, he wrote Liftoff: the Story of America's Adventure in Space.

Over the past decades, Moon hoax advocates have brought up many intriguing points regarding the Apollo missions to the moon. Believers claim that none of them, however, are indisputable evidence of fraud, and most are misunderstandings of basic physics or mission specifications.

Several lines of evidence point to an actual Moon landing. The 880 pounds of Moon rocks brought back to Earth

THE MOON LANDING HOAX

are the most obvious evidence, as well as the testimonies of 12 astronauts. More than 30,000 photographs were brought back from the Apollo missions, as well as hours upon hours of video footage.

Lunar reflectors, image courtesy NASA

The surest evidence claimed though, that men have walked on the moon is in lasers. Apollo astronauts placed a reflecting mirror on the moon for scientific purposes. If you point a laser at that precise point on the moon, the beam is reflected back. Scientists have used this method to measure the distance to the moon with amazing precision, and it is from such experiments that we know that the moon is slowly receding away from the earth.

Many hoax believers are well meaning people who have been duped into believing the hoax theories by what they perceive to be compelling evidence. There are other hoax advocates, often representing themselves as experts, who publicly make claims based on erroneous conclusions resulting from a lack of proper research, scientific ignorance, or extreme prejudice.

A third possibility is that there are those who may believe the moon landings were real, but intentionally try to persuade people otherwise for some sort of attention, fame or profit. But may be this time around they were true.

The problem the hoax advocates face is that there is a mountain of evidence supporting the authenticity of the moon landings. In order to substantiate their story, this evidence must be refuted. In some cases, the hoax advocates propose arguments that, on the surface, appear to have some merit, but as they try to dismiss other evidence it becomes more difficult. Usually their claims become more and more outlandish, often times foolish. In many cases they resort to making assertions that are seriously flawed in both science and logic. In short most of the claims even though it makes sense it is circumstantial.

PART 2

THE LANDING – NASA OFFICIAL REPORTS AND THE BEGINNING OF DOUBTS

Down to 50,000 feet above the Moon, Apollo 11 was little different from 10. Armstrong, Aldrin, and Command Module Pilot Mike Collins had a flawless launch from Earth, a long, uneventful coast out to the Moon, and a nominal engine burn to put themselves into lunar orbit. On that first orbit, the planned landing site itself was still enveloped in pre-dawn darkness and it wasn't until their fourth pass - during Lunar Module checkout that Aldrin reported seeing it from the Lunar Module windows.

Eighty six hours and five-and-a-half lunar orbits into the mission, the crew of Apollo 11 settled down for their last rest period before the landing. The final rest before the landing was necessarily a short one, but the three of them each got six hours of deep sleep.

For the next eight hours, Armstrong, Aldrin, and Collins got ready for the descent. By the time they disappeared behind the Moon for the fourteenth time, they were suited and undocked, and just minutes away from the 30-second burn that would put the Lunar Module on the Apollo 10 descent trajectory down to 50,000 feet. Collins would remain in a circular orbit 60 nautical miles above the Moon and, because of his higher altitude, it was he who was first to regain radio contact with Earth. Everything had gone well. The Lunar Module would be coming around the corner in just a moment and right on time.

There were no seats in the Lunar module. Armstrong and Aldrin were standing, held in place by elastic cords attached to the flooring. For sixteen minutes they looked out the windows and timed the passage of landmarks below them to confirm the tracking data that Houston was getting. With Houston's help, they also checked and double checked the health of the Lunar module.

When the twelve-and-a-half-minute descent burn finally began, they had the spacecraft oriented so that they were flying with their feet and the engine forward. They were also flying with the windows facing the Moon so that they could do some post-ignition landmark timing; but then, as planned, three minutes into the burn Armstrong rotated the spacecraft to a face-up position.

Now, he and Aldrin needed to fly with their backs to the Moon so that, as they approached the landing site and the LM began to rotate upright, Armstrong would be able to see the ground ahead and pick out a good, clear landing spot.

The Lunar module, image courtesy NASA

As they flew, they monitored the Lunar module's performance and the readouts of the guidance computer. All of the data indicated that they were flying very close to the planned trajectory.

The program alarms and the communications dropouts were annoying but in all other respects, the Lunar module computer and the navigation system performed beautifully. Eight minutes and thirty seconds into the burn, the computer pitched the Lunar module nearly upright and Armstrong got his first close-up view of the place to which the computer was taking them. He was about 5,000 feet above and about 20,000 feet east of it. As planned, he had fuel enough for five more minutes of flight. Each of the astronauts had a small, double-paned, triangular window in front of him.

On the inner surface of each pane in Armstrong's window, there was a long vertical scale marked in degrees and, at right angles to it, a similar but shorter horizontal scale. At pitch over, Armstrong positioned himself so that the vertical scales were aligned; and Aldrin read a computer output to him that indicated just where he should look on the scale to find the computer's intended landing point. In principle, if he didn't like the spot, he could pulse the pistol-grip hand controller forward or back or to either side and, thereby tell the computer to move the target a small amount in the indicated direction.

According to plan, Aldrin was to give Armstrong an "angle" every few seconds until, at an altitude of about five hundred feet, the window targeting lost its usefulness and Armstrong took over complete manual control for the final descent.

Even with the computers to help, landing a Lunar module was a tricky operation, one that required countless hours of training in indoor simulators and in an ungainly "flying bedstead" called - more formally - the Lunar Landing Training Vehicle.

The Lunar Landing Training Vehicle, image courtesy NASA

While Armstrong flew the Lunar module toward a good landing spot, his attention was totally focused on the job at hand. Aldrin did virtually all the talking; and he, too, was all business. He read the computer output to Armstrong, giving him their altitude, their rate of descent and their forward speed.

Back in Houston, Flight Director Gene Kranz and other members of the support team in the Mission Control Room were watching telemetry from the LM.

Apollo 11 Mission Control , image courtesy NASA

They did not know about the crater yet - Armstrong wouldn't discuss it until well after the landing - but it was obvious that the landing was taking longer than planned. Indeed, with each passing second there was mounting concern about how much fuel remained. Because of uncertainties in both the gauges in the tanks and the estimates that could be made from telemetry data on the engine firing, the amount of time remaining until the fuel ran out was uncertain by about 20 seconds. If they got too low, Kranz would have to order an abort.

Despite the drama of the moment and the enormous feelings of elation and relief that they both felt, Armstrong and Aldrin had little time for anything but getting the Lunar module ready for an immediate departure. No one expected that they would have to launch right away but, just in case a problem did develop - say a leak of the high pressure helium that they would use to pressurize the propellant tanks in the Ascent Stage, they wanted to be ready. However, despite being very busy with things in the spacecraft for nearly two hours after they landed, from time to time they stole glances out the window and described the scene for the radio audience back on Earth.

About six and a half hours after the landing, they had the hatch open and Armstrong crawled out onto the porch - feet first and on his hands and knees. Moments later he was on the top rung of the ladder and pulled a lanyard to release a workbench/stowage area that was attached to the side of the Lunar module. The Modular

Equipment Storage Assembly or MESA was pivoted at the bottom so that, when Armstrong pulled the lanyard, the MESA swung down into a horizontal position. The most important piece of gear on it was undoubtedly the black-and-white TV camera. It was mounted in such a way that, when the MESA swung down, the camera was pointed directly at the foot of the ladder. For the astronauts, the landing had been the big moment of the mission. But, for the waiting world, the big moment was still to come - the first footstep.

From the bottom rung of the ladder, Armstrong had to make a three-foot jump down to the footpad - a contingency against a less than gentle landing that might have compressed and shortened the landing strut. From the footpad, he had only a couple of inches to step down to the surface itself. He stood on the pad for a moment or two, testing the soil with the tip of his boot before he made the epochal "small step".

Aldrin joined Armstrong out on the surface about fifteen minutes later and then, for the next hour and forty minutes, the two of them examined the Lunar module, moved the TV camera out about 50 feet, deployed a pair of scientific instruments, and collected more samples.

With only a short time at their disposal, Armstrong and Aldrin had only a few things they could get done before they had to have the hatch closed. They raised an American Flag, deployed a solar wind collector, gathered forty-seven pounds of samples, and carried a laser reflector and the passive seismometer about twenty meters south of the Lunar module for deployment.

Neil Armstrong on the Lunar surface, image courtesy NASA

They hammered two, short core tubes into the soil, took about one hundred color photographs, and, finally, got themselves and the samples back into the spacecraft. Because of the restrictions imposed by the pressurized suits, by the shifted center of mass, by the weak gravitational field, and especially by the clumsy, pressurized gloves, the work was generally harder to do than it would have been in a shirtsleeve environment.

Two hours and thirty one minutes after they first opened the hatch, Armstrong and Aldrin reported it closed again. And returned to earth

The NASA report on the Apollo mission was quite elaborate, and well detailed, But was it real, Did man really walk on the Moon or was it the ultimate camera trick.

It is of course the conspiracy theory to end all conspiracy theories. The story lifts off in 1961 with Russia firing Yuri Gagarin into space, leaving a panicked America trailing in the space race. At an emergency meeting of Congress, President Kennedy proposed the ultimate face saver; put a man on the Moon. With an impassioned speech he secured the plan an unbelievable 40 billion dollars. And many claim thus, the great Moon hoax was born.

Between 1969 and 1972, seven Apollo ships headed to the Moon. Six claim to have made it, with the ill fated Apollo 13--whose oxygen tanks apparently exploded halfway--being the only casualties.

But with the exception of the known rocks, which could have been easily mocked up in a lab, the photographs and film footage are the only proof that the Eagle ever landed. The world tuned in to watch what looked like two blurred white ghosts gambol threw rocks and dust. Part of the reason for the low quality was that, strangely, NASA provided no direct link up. So networks actually had to film "man's greatest achievement" from a TV screen in Houston--a deliberate ploy, some say, so that nobody could properly examine it. By contrast, the still photos were stunning. Yet that's just the problem. The astronauts took thousands of pictures, each one perfectly exposed and sharply focused. Not one was badly composed or even blurred.

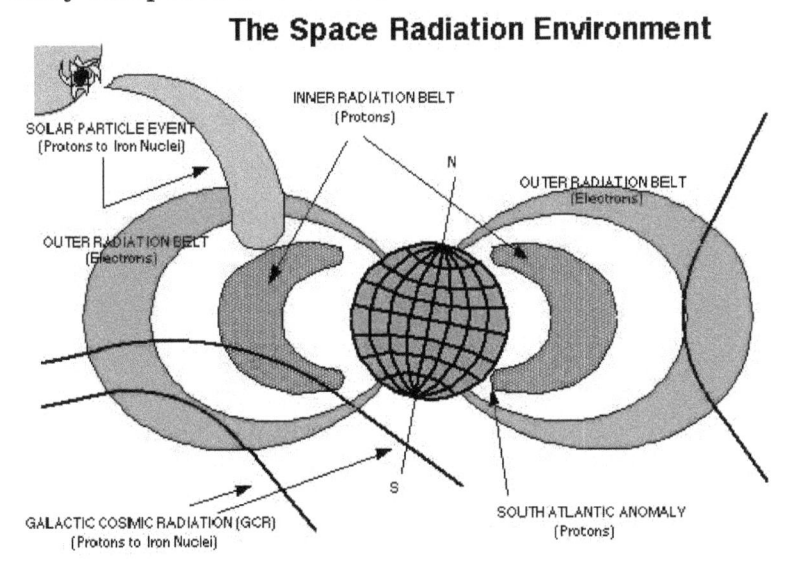

The Space Radiation Environment

Representation of the major sources of ionizing radiation of importance to manned missions in low-Earth orbit. Note the spatial distribution of the trapped radiation belts.

Space radiation environment, image courtesy NASA

The questions don't stop there. Outer space is awash

THE MOON LANDING HOAX

with deadly radiation that emanates from solar flares firing out from the sun. Standard astronauts orbiting earth in near space, like those who recently fixed the Hubble telescope, are protected by the earth's Van Allen belt. But the Moon is 240,000 miles distant, way outside this safe band. And, during the Apollo flights, astronomical data shows there were no less than 1,485 such flares.

Every Apollo mission before number 11 was plagued with around 20,000 defects a-piece. Yet, with the exception of Apollo 13, NASA claims there wasn't one major technical problem on any of their Moon missions. Just one defect could have blown the whole thing.

Several years after NASA claimed its first Moon landing, Buzz Aldrin "the second man on the Moon"--was asked at a banquet what it felt like to step on to the lunar surface. Aldrin staggered to his feet and left the room crying uncontrollably. Virgil Grissom, a NASA astronaut, was due to pilot Apollo 1. In January 1967, he baited the Apollo program by hanging a lemon on his Apollo capsule and told his wife Betty: "if there is ever a serious accident in the space program, it's likely to be me."

Nobody knows what fuelled his fears, but by the end of the month he and his two co-pilots were dead, burnt to death during a test run when their capsule, pumped full of high pressure pure oxygen, exploded. Scientists couldn't believe NASA's carelessness--even a chemistry

PART 2

student in high school knows high pressure oxygen is extremely explosive. In fact, before the first manned Apollo fight even cleared the launch pad, a total of 11 would be astronauts were dead. Apart from the three who were incinerated, seven died in plane crashes and one in a car smash. Now this is a spectacular accident rate.

One wonders if these 'accidents' weren't NASA's way of correcting mistakes, of saying that some of these men didn't have the sort of 'right stuff' they were looking for.

PHOTOGRAPHIC EVIDENCE FOR AND AGAINST
The cameras had no white meters or view finders. So the astronauts took photographs without being able to see what they were doing. These photographs were in the midst of the controversies as There film stock was unaffected by the intense peaks and powerful cosmic radiation on the Moon, conditions that should have made it useless. They managed to adjust their cameras, change film and swap filters in pressurized clubs. It should have been almost impossible to bend their fingers. . The shadows could only have been created with multiple light sources and, in particular, powerful spotlights. But the only light source on the Moon was the sun. The American flag and the words "United States" are always brightly lit, even when everything around is in shadow.

Not one still picture matches the film footage, yet NASA claims both were shot at the same time. The pictures are so perfect; each one would have taken a slick advertising agency hours to put them together. But the astronauts managed it repeatedly.

Apollo 14 astronaut Allen Shepard played golf on the Moon. In front of a worldwide TV audience, Mission Control teased him about slicing the ball to the right. Yet a slice is caused by uneven air flow over the ball.

Alan Shepard 'plays' golf, image courtesy NASA

The Moon has no atmosphere and no air. A camera panned upwards to catch Apollo 16's Lunar Lander lifting off the Moon. Who did the filming? One NASA picture

PART 2

from Apollo 11 is looking up at Neil Armstrong about to take his giant step for mankind. The photographer must have been lying on the planet surface. If Armstrong was the first man on the Moon, then who took the shot? The pressure inside a space suit was greater than inside a football, but were seen freely bending their joints.

Why didn't America make a signal on the Moon that could be seen from Earth? The PR would have been phenomenal and it could have been easily done with magnesium flares. Text from pictures in the article show only two men walked on the Moon during the Apollo 12 mission. Yet the astronaut reflected in the visor has no camera. Who took the shot? And why is the flag fluttering?

Visor reflection, image courtesy NASA

THE MOON LANDING HOAX

How can the flag be brightly lit when its not facing any light ? And where, in all of these shots, are the stars?

Brightly lit and fluttering flag, image courtesy NASA

The Lander weighed 17 tons yet the astronauts feet seem to have made a bigger dent in the dust. The powerful booster rocket at the base of the Lunar Lander was fired to slow descent to the moons service. Yet it has left no traces of blasting on the dust underneath. It should have created a small crater, yet the booster looks like it's never been fired...

But the biggest shock is yet to come! The camera pans left past Neil Armstrong towards the left hand side of the Apollo 11, there seems to be what appears to be another

Earth. It must also be noted that the Apollo 11 at this point of the mission was supposedly half way to the Moon. The time elapsed was 34 hours and 16 minutes, but from the view of Earth in the right hand window, we can say that in fact they were not in deep space at all, but still in low Earth orbit! look at the blue sky outside.

That would also explain why they would be filming an exposure of the Earth that was far away, to give the impression that they were in deep space. The exposure would be clipped to the window and the Sun's luminance would light it up, a technique that was used to read star charts to help with navigation and star reference.

An important factor to take into consideration is the great variations in temperature that the film would have had to endure whilst on the lunar surface. The temperature during the Apollo missions were recorded as being between -180 Fahrenheit in the shade to an incredible +200 Fahrenheit in full Sunshine. How could the film emulsion have withstood such temperature differences? The astronauts can be seen to move between the shadows of the rocks and then into full sunlight in some shots. Surely the film would have perished under such conditions? One of the biggest anomalies that appear on the Moon shots are the way in which shadows seem to be cast in totally different directions, even when the objects making the shadows are a mere few feet apart?

The lunar lander used two engines stacked on top of one another. The Lunar module's descent engine used hyperbolic propellants, that means two different fuels that light at the same time. The exhaust jet coming out of the LEM on descent or ascent should have created an enormous cloud of reddish coloured gas, instead we see the bursting apart of the similar covering as it leaves the Moons surface? The fuel used are exactly the same as used on the Shuttle today, and we can clearly see the exhaust smoke coming from them, so why not the Lunar module?

No crater below Lunar Lander, image courtesy NASA

Surely there should have been some type of crater under the Apollo landing modules, especially the Apollo 12, as it slowly moved across the moon's surface before landing. The 5000 degree Fahrenheit heat from the 10,000 lb thrust of the engine should have produced at least some volcanic rock. If you compare the molten volcanic rock at Mount Etna, that was boiled at only 1000 Celsius. Some skeptics claim that the engines force would have been dispersed mainly sideways, but if this is so, what actually held up the 2,300lbs of lunar lander when it was on its descent to the lunar surface? Why was there not any dust in the landing pads either? There is certainly lots of dust scattered when the Lunar module is leaving the Moon and if the engine simply blew all the dust away from around the Lunar module as it landed, how did Armstrong manage to create that famous footprint?

FACTS ABOUT THE MOON.
The Moon is the only natural satellite of Earth, and it orbits 384,400 km from Earth. The moon has a diameter of 3476 km with a mass of 7.35e22 kg. The moon was called Called Luna by the Romans, Selene and Artemis by the Greeks, and many other names in other mythologies.

The Moon, of course, has been known since prehistoric times. It is the second brightest object in the sky after the Sun. As the Moon orbits around the Earth once per month, the angle between the Earth, the Moon and the Sun changes; we see this as the cycle of the Moon's phases. The time between successive new moons is 29.5 days i.e 709 hours, slightly different from the Moon's orbital period since the Earth moves a significant distance in its orbit around the Sun in that time.

Due to its size and composition, the Moon is sometimes classified as a terrestrial "planet" along with Mercury, Venus, Earth and Mars.

Full moon, image courtesy NASA

The Moon was first visited by the Soviet spacecraft Luna 2 in 1959. It is the only extraterrestrial body to have been visited by humans. The first landing was on July 20, 1969 the last was in December 1972. The Moon is also the only body from which samples have been returned to Earth. In the summer of 1994, the Moon was very extensively mapped by the little spacecraft Clementine and again in 1999 by Lunar Prospector.

The 1999 Lunar Prospector, image courtesy NASA

The gravitational forces between the Earth and the Moon cause some interesting effects. The most obvious is the tides. The Moon's gravitational attraction is stronger on the side of the Earth nearest to the Moon and weaker on the opposite side. Since the Earth, and particularly the oceans, is not perfectly rigid it is stretched out along the line toward the Moon. From our perspective on the Earth's surface we see two small bulges, one in the direction of the Moon and one directly opposite. The effect is much stronger in the ocean water than in the solid crust so the water bulges are higher. And because the Earth rotates much faster than the Moon moves in its orbit, the bulges move around the Earth about once a day giving two high tides per day.

But the Earth is not completely fluid, either. The Earth's rotation carries the Earth's bulges slightly ahead of the point directly beneath the Moon. This means that the force between the Earth and the Moon is not exactly along the line between their centers producing a torque on the Earth and an accelerating force on the Moon. This causes a net transfer of rotational energy from the Earth to the Moon, slowing down the Earth's rotation by about 1.5 milliseconds/century and raising the Moon into a higher orbit by about 3.8 centimeters per year.

The asymmetric nature of this gravitational interaction is also responsible for the fact that the Moon rotates synchronously, i.e. it is locked in phase with its orbit

PART 2

so that the same side is always facing toward the Earth. Just as the Earth's rotation is now being slowed by the Moon's influence so in the distant past the Moon's rotation was slowed by the action of the Earth, but in that case the effect was much stronger. When the Moon's rotation rate was slowed to match its orbital period there was no longer an off-center torque on the Moon and a stable situation was achieved. The same thing has happened to most of the other satellites in the solar system. Eventually, the Earth's rotation will be slowed to match the Moon's period, too, as is the case with Pluto and Charon.

Actually, the Moon appears to wobble a bit so that a few degrees of the far side can be seen from time to time, but the majority of the far side) was completely unknown until the Soviet spacecraft Luna 3 photographed it in 1959. There is no "dark side" of the Moon; all parts of the Moon get sunlight half the time except for a few deep craters near the poles. Some uses of the term "dark side" in the past may have referred to the far side as "dark" in the sense of "unknown" but even that meaning is no longer valid today

The Moon has no atmosphere. But evidence from Clementine suggested that there may be water ice in some deep craters near the Moon's south pole which are permanently shaded. This has now been reinforced by data from Lunar Prospector. There is apparently ice at the north pole as well. A final determination will probably come from NASA's Lunar Reconnaissance Orbiter, scheduled for 2008.

The Moon's crust averages 68 km thick and varies from essentially 0 under Mare Crisium to 107 km north of the crater Korolev on the lunar far side. Below the crust is a mantle and probably a small core roughly 340 km radius and 2% of the Moon's mass. Unlike the Earth, however, the Moon's interior is no longer active. Curiously, the Moon's center of mass is offset from its geometric center by about 2 km in the direction toward the Earth. Also, the crust is thinner on the near side.

There are two primary types of terrain on the Moon: the heavily cratered and very old highlands and the relatively smooth and younger maria. The maria which comprise about 16% of the Moon's surface are huge impact craters that were later flooded by molten lava. Most of the surface is covered with regolith, a mixture of fine dust and rocky debris produced by meteor impacts. For some unknown reason, the maria are concentrated on the near side.

Most of the craters on the near side are named for famous figures in the history of science such as Tycho, Copernicus, and Ptolemaeus. Features on the far side have more modern references such as Apollo, Gagarin and Korolev with a distinctly Russian bias since the first images were obtained by Luna 3. In addition to the familiar features on the near side, the Moon also has the huge craters South Pole-Aitken on the far side which is 2250 km in diameter and 12 km deep making it the the largest impact basin in the solar system and Orientale on the western limb.

A total of 382 kg of rock samples were returned to the Earth by the Apollo and Luna programs. These provide most of our detailed knowledge of the Moon. They are particularly valuable in that they can be dated. Even today, more than 30 years after the last Moon landing, scientists still study these precious samples.

Most rocks on the surface of the Moon seem to be between 4.6 and 3 billion years old. This is a fortuitous match with the oldest terrestrial rocks which are rarely more than 3 billion years old. Thus the Moon provides evidence about the early history of the Solar System not available on the Earth.

Prior to the study of the Apollo samples, there was no consensus about the origin of the Moon. There were three principal theories: co-accretion which asserted that the Moon and the Earth formed at the same time from the Solar Nebula; fission which asserted that the Moon split off of the Earth; and capture which held that the Moon formed elsewhere and was subsequently captured by the Earth. None of these work very well. But the new and detailed information from the Moon rocks led to the impact theory: that the Earth collided with a very large object as big as Mars or more and that the Moon formed from the ejected material. There are still details to be worked out, but the impact theory is now widely accepted.

The Moon has no global magnetic field. But some of its surface rocks exhibit remanent magnetism indicating that there may have been a global magnetic field early in the Moon's history.

With no atmosphere and no magnetic field, the Moon's surface is exposed directly to the solar wind. Over its 4 billion year lifetime many ions from the solar wind have become embedded in the Moon's regolith. Thus samples of regolith returned by the Apollo missions proved valuable in studies of the solar wind.

MOTIVES TO FAKE THE MOON LANDING

The main reason why the US Government and NASA faked the 'official record' is because they could not be seen to be the weak link, especially when you consider that during the 60's, the USA were at the height of the Cold War with Russia. Also their own President had forecast that before the end of the 60's Man would be on the Moon. It would be better to try and fool the public and hoax the footage, rather than let their biggest rival in the World strike a huge moral victory by beating them to the Moon.

Several motives have been suggested for the U.S. government to fake the moon landings

The U.S. government benefited from a popular distraction to take attention away from the Vietnam War. The Vietnam War was getting a lot of criticism from home and the attention had to be diverted. Lunar activities did abruptly stop, with planned missions cancelled, around the same time that the US ceased its involvement in the Vietnam War. The U.S. government considered it vital that the U.S. win the space race with the USSR. Going to the Moon, if it was possible, would have been risky and expensive. It would have been much easier to fake the landing, thereby ensuring success.

NASA raised approximately 30 billion dollars pretending to go to the moon. This could have been used to pay off a large number of people, providing significant motivation for complicity. In variations of this theory, the space industry is characterized as a political economy, much like the military industrial complex, creating fertile ground for its own survival.

Another reason for faking the landing might be because the available technology at the time was such that there was a good chance that the landing might fail if genuinely attempted. Writer Bill Kaysing is one of the prime movers in the conspiracy said that NASA couldn't make it to the Moon, and they knew it... In the late -50s, when he was at Rocketdyne, they did a feasibility study on astronauts landing on the Moon. They found that the chance of success was something like .0017 percent. In other words, it was hopeless.- As late as 1967, Kaysing says, -three astronauts died in a horrendous fire on the launch pad. It's also well documented that NASA was often badly managed and had poor quality control. But as of 1969, we could suddenly perform manned flight upon manned flight? With complete success? It was just against all statistical odds.

But President John F. Kennedy publicly announced in May 1961 that -landing a man on the Moon and returning him safely to earth- would be a Number One priority for the US, an accomplishment that was to instill pride

PART 2

in Americans and awe in the rest of the world. Hence failure was not an option. No one would argue that the mission succeeded.

President Kennedy speaks to Congress, 1961
image courtesy NASA

For JFK and his government, image was everything. The relevance to WYSINWYSHG is obvious, but the consequences on popular culture were, and still are, huge. For starters, this was a watershed in the power of TV. -If television ever had a killer app, the Moon landing was it. The Americans bought new sets in droves, flicked them on as zero hour approached, and, miraculously, felt them being locked into an intangible but very real oneness with a billion other people. It was our first taste of a virtual community, of cultures docking. It felt good, and it was essential for a country in crisis. Watergate hadn't happened yet, and people still trusted their elected officials. The Cold War, with the unseen, faceless enemy, made politicians at home seem a lot more believable people are easier to lead when there is an external enemy to unite against. After Vietnam and Watergate, people have become less trusting, and to some people it doesn't matter what the government says.

The Apollo missions and space exploration in general elevated science to God status. The repercussions of this are still being felt globally. The cold war was not only about consumerism vs. Communism, it was a war of science. How could it all have been a fake? How could NASA possibly have pulled it off? What about the TV pictures that billions of people saw over the course of six successful missions; the rocket lifting off from the Cape Kennedy launch pad under the watchful eye of hundreds of thousands of spectators; the capsule with the crew returning to earth; the Moon rocks; the hundreds of space-program employees in the know who would have to be relied upon to take the incredible secret to their graves? If it was faked no one would believe it.

PART 2

The Moon landings bred a culture of science in the -60s. Anything would be possible with the new God - that of science. Science would one day render nuclear reactors safe, we would holiday on Mars, and colonise the solar system. The belief of the Moon landings and our superior technology also took away flak from the growing anti-bomb movement - this was positive science.

Saturn V rocket on the launch pad, image courtesy NASA

NASA had good reason to stage Moon landing after Moon landing - both NASA and Rocketdyne wanted the money to keep pouring in. Exploration of the Moon stopped because it was impossible to continue the hoax without being ultimately discovered. And of course they ran out of pre-filmed episodes. If man has set foot on the Moon, then we have not seen the images, and the chances are, if he returned to this planet he would not have been able to tell the tale.

There were six Moon landings costing upwards of US $25 billion. Proving to the world that not even the Soviets could come close to the US when it came to space exploration. After a slow start, the Americans won the space race. Money was also another key factor, 25 Billion Dollars of the tax payers money was a huge amount, and success had to achieve at all cost even if it meant faking the whole mission.

THE SOVIET UNION – COMPETITION IN SPACE, THE COLD WAR, VIETNAM

The Cold War began after World War Two. The main enemies were the United States and the Soviet Union. The Cold war got its name because both sides were afraid of fighting each other directly. In such a "hot war," nuclear weapons might destroy everything. So, instead, they fought each other indirectly. They supported conflicts in different parts of the world. They also used words as weapons. They threatened and denounced each other. Or they tried to make each other look foolish.

The United States and the Soviet Union were the only two superpowers following the Second World War. The fact that, by the 1950s, each possessed nuclear weapons and the means of delivering such weapons on their enemies, added a dangerous aspect to the Cold War. The Cold War world was separated into three groups. The United States led the West. This group included countries with democratic political systems. The Soviet Union led the East. This group included countries with communist political systems. The non-aligned group included countries that did not want to be tied to either the West or the East.

During the Second World War, the Soviet Union was an ally of the Western democracies, in their struggle against the Axis Powers of Germany, Japan and Italy. As the War neared its conclusion, the future of Eastern Europe became a point of contention between the Soviet Union and its Western allies. The Soviet Union had been invaded via Eastern Europe in both the First and Second World Wars. In both conflicts, some of the nations of Eastern Europe had participated in those invasions. Both Wars had devastated the Soviet Union. An estimated twenty-five million Russians were killed during the Second World War. The Soviet Union was determined to install "friendly" regimes throughout Eastern Europe following the War. The strategic goal was to protect its European borders from future invasions. Since the Soviet Union was a communist state, the Soviet government preferred to install communist regimes throughout Eastern Europe. The Red Army was liberating the nations of Eastern Europe and therefore, the Soviet Union was in a position to influence the type of governments that would emerge following the War.

The Western democracies, led by the United States, were determined to stop the spread of communism and Soviet power. While not being able to stop the Soviets in Eastern Europe, the U.S. and Britain were determined to prevent communist regimes from achieving power in Western Europe. During the Second World War, communists parties throughout Western Europe, had gained popularity in their resistance to Nazi occupation. There was a real possibility the communist parties would be elected in both France and Italy.

Harry Truman was the first American president to fight the Cold War. He used several policies. One was the Truman Doctrine. This was a plan to give money and military aid to countries threatened by communism. The Truman Doctrine effectively stopped communists from taking control of Greece and Turkey. Another policy was the Marshall Plan, which provided financial and economic assistance to the nations of Western Europe.

This strengthened the economies and governments of countries in western Europe, and as the economies of Western Europe improved, the popularity of communist parties declined.

The United States also led the formation of the North Atlantic Treaty Organization in 1949. NATO was a joint military group. Its purpose was to defend against Soviet forces in Europe or, as the saying went, "to keep Russia out, America in and Germany down". The first members of NATO were Belgium, Britain, Canada, Denmark, France, Iceland, Italy, Luxembourg, the Netherlands, Portugal, and the United States. The Soviet Union and its east European allies formed their own joint military group -- the Warsaw Pact -- six years later.

The United States and the USSR both armed themselves and prepared for being attacked by the other country. Each country spent billions of dollars and raced to develop more effective military weapons, such as the atomic bomb, than the other. The Arms Race was a period of weaponary development in the world. Both countries strived to develop a stronger nuclear bomb. Also, nuclear rivalry led to the invention of a long line of increasingly deadlier weapons. Because of the Arms Race, both countries have to many nuclear warheads and must pay billions of dollars to have them destroyed. During the Cold War the United States and Russia didn't only compete for nuclear weapons, they also competed in

THE MOON LANDING HOAX

space. In 1957, the Russia sent Sputnick, a satellite, into orbit and caused America to worry because no one knew what the satellite was capable of doing. When Congress heard of this, it formed NASA, National Aeronautics and Space Administration, to develop America's technology. America then sent their own satellite into outer space, almost as a reply to Russia. Although Russia sent the first man into space in 1961, the United States had the first man on the moon in 1969. Both countries tried to develop their spy technology by using satellites in space.

Cold War tensions increased, then eased, then increased again over the years. The changes came as both sides actively tried to influence political and economic developments around the world. For example, the Soviet Union provided military, economic, and technical aid to communist governments in Asia. The United States then helped eight Asian nations fight communism by establishing the Southeast Asia Treaty Organization. In the middle 1950s, the United States began sending military advisers to help South Vietnam defend itself against communist North Vietnam. That aid would later expand into a long and bloody period of American involvement in Vietnam.

With the cold war at its height, the US leaders were worried that an attack on North Vietnam by the US would create tensions with the Chinese and Russians that would, in turn, lead to a larger conflict and possibly

WW III. This created a difficult situation for the US and would eventually lead to many internal conflicts, which ultimately prevented the US from forming a firm policy for the region. The US was also faced with a number of cultural differences between the two countries, and what was considered corrupt by the US government was considered legitimate by South Vietnamese standards. It was difficult for the US to portray South Vietnam as a hard working, hard fighting democracy; corruption was widespread among officials and the armed forces. The Army of the Republic of Vietnam was disorganized due to the low morale of it's leaders and their singular interest in personal gain. Therefore the US had a great deal of difficulty in holding the army together in South Vietnam and saw only one solution, that was to start taking care of things for themselves. By 1950 the US began sending their first troops, firstly in an advisory role, which slowly escalated into a full blown commitment.

The large-scale involvement of the US came under the tenure of President Lyndon B. Johnson and his Gulf of Tonkin Resolution. Johnson had replaced John F. Kennedy after he was assassinated in Dallas, Texas 1963. As president, he was torn between the differing strategies the US had for Vietnam. The increasing involvement and the escalation of troop involvement meant there were more casualties and more problems at home. But Johnson, who was always concerned about his image, and as president, held the power to halt the war in Vietnam, could not face the thought of being regarded as the first president in US history to loose a war.

The pressure around him grew so intense, that he was only left with one option and that was not to run for a second term. Basically, he handed the hot potato to Richard M. Nixon.

The top US commander in Vietnam was General William Westmoreland; he had to face an army full of young men placed in an environment that was totally alien to them. There was no clear front to the conflict and basically, the enemy could be hiding anywhere and everywhere. Life in the jungle was tortuous and there were no home comforts. Drugs and other stimulants filtered their way into the daily routine of many servicemen and morale quickly started to fall. For the first time, people in the US resisting the draft were given acceptance although still not by the majority of citizens. Riots and demonstrations against the war became the norm in the US, with numerous veterans taking part in the efforts to stop the war, strengthening the issue. Finally, the US government saw that it was in a no-win situation and began making plans to withdraw.

After great efforts by the US to withdraw, and the establishment of a cease-fire on January 27th, 1973, American soldiers began leaving Vietnam for good. The

North Vietnamese finally conquered South Vietnam in early 1975, totally ignoring the cease-fire and on July 2nd, 1976, North and South Vietnam were officially united as a single communist state. It had cost an estimated 2 million lives and the injury or disablement of many millions of others.

THE MOON LANDING HOAX

PART 3

Strange deaths of key people involved with the Apollo program

Some of the Eleven Apollo astronauts had non space related fatal accidents within a twenty two month period of one another, the odds of this happening are 1 in 10,000...coincidence.

Along with this key members of the Apollo program resigned from NASA, in quick succession, James B. Irwin of Apollo 15 resigned from NASA and the Air Force on July 1, 1972. Don F. Eisele of Apollo 7 resigned from NASA and from the Air Force in June 1972. Stewart Allen Roosa of Apollo 14 resigned from NASA and retired from the Air Force in February 1976. Swigert resigned from NASA in 1977. Why did they all resign from the 'successful' Apollo Program?

The program hints darkly that a number of astronauts and other witnesses were done away with to protect the secret. For example, ten astronaut trainees, 15% of the total, died in accidents between 1964 and 1967. Well, being a jet pilot is risky business.

Then there's the mysterious fire that killed Gus Grissom, Ed White and Roger Chaffee in 1967.

Remains of Apollo 1, image courtesy NASA

Was it due to Grissom's being a thorn in the side of NASA? Well, if so, there was a much neater and less messy solution to the problem. Ground him. There was more than enough reason - Grissom lost his capsule on his first flight! When Scott Carpenter overshot his landing site by 300 miles on the second Mercury orbital flight, he vanished from the space program. Somehow Grissom redeemed himself well enough to get a second chance aboard a Gemini mission, and he was set to become America's first three-time space traveler when he was killed.

In a television program about the hoax theory, Fox Entertainment Group listed the deaths of 10 astronauts and of two civilians related to the manned spaceflight program as having possibly been killings as part of a cover-up. Ted Freeman died in the T-38 crash in1964. Elliott See and Charlie Bassett died in the T-38 accident in 1966. Virgil "Gus" Grissom supposedly an outspoken critic of the Space Program died in the Apollo 1 fire, January 1967, Ed White & Roger Chaffee died in the Apollo 1 fire, January 1967, Ed Givens died in an car accident, 1967, C. C. Williams died in the T-38 accident in October 1967, X-15 pilot Mike Adams was the only X-15 pilot killed in November 1967 during the X-15 flight test program - not a NASA astronaut, but had flown X-15 above 50 miles. Robert Lawrence, scheduled to be an Air Force Manned Orbiting Laboratory pilot who died in a jet crash in December 1967, shortly after reporting for duty to that later cancelled program. NASA worker Thomas Baron train crash, 1967 shortly after making accusations before Congress about the cause of the Apollo 1 fire, after which he was fired. The death was ruled as a suicide. James Irwin, an Apollo 15 astronaut who was referred to as an "informant", to confess about a cover-up having occurred.

James Irwin, image courtesy NASA

Irwin was supposedly going to contact a journalist about it; however he died of a heart attack in 1991, before any such telephone call occurred.

Landing believers explain these deaths by pointing out that spacecraft testing and flying high performance jet aircraft can be dangerous, and all but one of the astronaut deaths were directly related to their rather hazardous job. Two of the astronauts, Mike Adams and Robert Lawrence, had no connection with the civilian manned space program. Astronaut James Irwin had suffered several heart attacks in the years prior to his death. It was strange and with skeptics questioning each and every aspect, deaths related to people associated with NASA became a part of the conspiracy. Whether they were silenced or fate no one would never know. These deaths rose questions that NASA couldn't give reply for and as a result arouse doubts as well as the skeptics to question the authenticity of the Apollo landings

The vast costs of the Apollo program – did success have to be achieved at all cost?

The US won the race to the Moon because, unlike the Soviet Union, it committed vast resources to a well thought-out "game plan" right from the start. NASA also stuck to that plan despite occasional technical and political problems. The foundation for Apollo's success was laid in 1962-67 when some 500,000 people from 20,000 companies built the spacecraft, Saturn carrier rocket and launch facilities. After this, the program was rapidly dismantled in just five years while the Apollo/Saturn system became operational, achieving President Kennedy's goal in July 1969 when Neil Armstrong became the first man on the Moon.

The cost of the entire Apollo program was USD $25.4 billion in1969 which would work out to $135-billion in 2005 Dollars. Apollo spacecraft and Saturn rocket cost alone, was about $ 83-billion 2005 Dollars

Project Mercury: This lasted from August 1959 to May 1963. The program was conceived in 1957 and consisted of Six manned flights, and it cost them approx. $1.5 billion ,Gemini Project started in 1962. Two unmanned, ten manned flights between April 1964 and November 1966, plus seven Gemini-Agena Target Vehicle launchings out of which five were successful. Although estimated to cost them 1.9 Billion dollars, the actual cost came up to 5.4 Billion dollars.

The Ranger which was the lunar impact and imaging missions consisted of nine probes including three fully successful missions launched between 1961 and 1965. The total cost worked up to approx. $1 billion.

Surveyor used for lunar soft landing was contracted to Hughes in 1961. Total estimated development cost was $300 million for1st mission in 1963, actual total cost worked out to: $2.8 billion, with seven missions between 1966 and 1968 out of which five were successful.

Lunar Orbiter Program started in 1963. There were five successful missions between 1966 & 1967. The total cost worked out to approx. $800 million.

Atlas and Titan were used in the Mercury and Gemini programs. The US Air Force's Agena upper stage was modified to serve as a docking target for Gemini, and also provided the 'final push' out of low Earth orbit for the Ranger and Lunar Orbiter probes. The major non-Apollo/Saturn related project was the Centaur cryogenic upper stage, started in 1959 for use in the Surveyor and Mariner programs. Eight Atlas-Centaur (AC) test flights between May 1962 and April 1966, including six total/partial failures. Expected cost of the project was less tha a billion, but they had to shell out close to 4 billion dollars for the project

Saturn series: Project started in 1960, initially comprising four different launchers. A fifth heavy-lift variant called 'Saturn C5' was chosen for Apollo in late 1961. Ten Saturn I launchings including four single-stage ballistic tests and three Pegasus satellites between 1961and 1965. Nine Saturn-IB missions between 1966 and 1975 including three Skylab, one Apollo-Soyuz. Thirteen Saturn V launchings between 1967 and 1973 including one Skylab.total cost of the project came up to $35 billion.

Work began on the Apollo CSM in November 1961, when NASA selected North American as main contractor. Two 'boilerplates' were launched in 1964. Two Block-I CSM prototypes were launched on ballistic test flights in 1966;

two more unmanned Block-Is flew on Saturn Vs in 1967/68. Fifteen manned Block-II spacecraft were launched in 1968-75, including three Skylab and one Apollo-Soyuz Test Project. the total cost came $17.5 billion.

The luna rmodule, image courtesy NASA

The Apollo Lunar Module was conceived in June 1962 when NASA decided to use the lunar orbit rendezvous technique rather than land the CSM on the Moon. Grumman won the contract in September 1962. The first unmanned tests took place in Earth orbit in 1968. Nine manned Lunar Module's were launched in 1969-72. NASA had to shell out 11 Billion for the project

Sky Lab station, image courtesy NASA

Although not part of the lunar program, the Skylab space station was nevertheless based on surplus Apollo hardware. The Skylab 1 laboratory cost about $7 billion, while the total cost of the three Apollo/Saturn IB flights to the station probably cost approx. $2 billion. Although NASA constructed two Skylabs, it could afford to launch only one of them.

Launching the second would have cost only $1.1 billion, plus $1.3 billion for two 2-month Apollo missions in 1974-76.

Project Apollo in general and the flight of Apollo 11 in particular, should be viewed as a watershed in American history. It was an endeavor that demonstrated both the technological and economic virtuosity of the United States and established national preeminence over rival nations— the primary goal of the program when first envisioned by the Kennedy administration in 1961. It had been an enormous undertaking, costing $25.4 billion i.e about $95 billion in 1990 dollars, with only the building of the Panama Canal rivaling the Apollo program's size as the largest non-military technological endeavor ever undertaken by the United States and only the Manhattan Project being comparable in a wartime setting.

With all these cost involved, NASA was under great pressure to show results. It wouldn't have been easy when the president promised an American on the moon by the end of the decade, under the cloud of a cold war, and trying to distract the countrymen from the disastrous Vietnam War. With the money pouring in for the ultimate face saver, success needed to be achieved at all cost.

Conclusions – was the Apollo 11 moon landing real or fake?

Like any long lasted argument it is extremely difficult to conclude what really happened, a mission of this magnitude would not be easy to manipulate. The whole mission involved close to 40.000 people and it would be extremely difficult to keep every body quite. If at all it was found out that what happened was a hoax, there it would have been a greater shame for the government. They would have to take great risk in even attempting such a thing. A considerable amount would have been needed to silence every one in the project.

But when we look at the situation and circumstances it evokes a question some where. NASA has not been straight forward or truthful with their answers. There is a whole lot of circumstantial evidence to prove that the government along with NASA took the great risk to save their own face. If at all such a thing had to happen it had to take place in America and in such a situation. The entire nation was coming to grips with the Vietnam war, as well as cold war. The political parties needed a saving grace to survive, and they needed something of this magnitude to help them.

The photographic evidence is very convincing, and the questions have not been answered. Moon being an alien atmosphere, it is difficult to formulate a solid evidence; If at all the conspiracies have to be answered we need another committee equipped with latest technology and enough money to prove them. But who can ask these questions; if the government was capable to spend time and money to silence people then they could do the same even now. The skeptics were far too small in numbers and without proper back up. To survive such a hoax for over 30 years, without a single whistle blower would be an achievement on its own.

Over the past decades, Moon hoax advocates have brought up many intriguing points regarding the Apollo missions to the moon. Believers claim that none of them, however, are indisputable evidence of fraud, and most are misunderstandings of basic physics or mission specifications.

Several lines of evidence point to an actual Moon landing. The 880 pounds of Moon rocks brought back to Earth are the most obvious evidence, as well as the testimonies of 12 astronauts. More than 30,000 photographs were brought back from the Apollo missions, as well as hours upon hours of video footage.

The surest evidence claimed though, that men have walked on the moon is in lasers. Apollo astronauts placed a reflecting mirror on the moon for scientific purposes. If you point a laser at that precise point on the moon, the beam is reflected back. Scientists have used this method to measure the distance to the moon with amazing precision, and it is from such experiments that we know that the moon is slowly receding away from the earth.

Many hoax believers are well meaning people who have been duped into believing the hoax theories by what they perceive to be compelling evidence. There are other hoax advocates, often representing themselves as experts, who publicly make claims based on erroneous conclusions resulting from a lack of proper research, scientific ignorance, or extreme prejudice. A third possibility is that there are those who may believe the moon landings were real, but intentionally try to persuade people otherwise for some sort of attention, fame or profit. But may be this time around they were true.

The problem the hoax advocates face is that there is a mountain of evidence supporting the authenticity of the moon landings. In order to substantiate their story, this evidence must be refuted. In some cases, the hoax advocates propose arguments that, on the surface, appear to have some merit, but as they try to dismiss other evidence it becomes more difficult. Usually their claims become more and more outlandish, often times foolish. In many cases they resort to making assertions that are seriously flawed in both science and logic. In short most of the claims even though it makes sense it is circumstantial.

SPACE TRAVEL GLOSSARY

A

a, A -- Acceleration. a = Δ velocity / Δ time. Acceleration = Force / Mass

A -- Ampere, the SI base unit of electric current.

Å -- Angstrom (0.0001 micrometer, 0.1 nm).

A Ring -- The outermost of the three rings of Saturn that are easily seen in a small telescope.

AAAS -- American Association for the Advancement of Science.

AACS -- Attitude and Articulation Control Subsystem onboard a spacecraft.

AAS -- American Astronomical Society.

AC -- Alternating current.

Acceleration -- Change in velocity. Note that since velocity comprises both direction and magnitude (speed), a change in either direction or speed constitutes acceleration.

ALT -- Altitude.

ALT -- Altimetry data.

AM -- Ante meridiem (Latin: before midday), morning.

am -- Attometer (10-18 m).

AMMOS -- Advanced Multimission Operations System.

THE MOON LANDING HOAX

Amor -- A class of Earth-crossing asteroid.

AO -- Announcement of Opportunity.

AOS -- Acquisition Of Signal, used in DSN operations.

Aphelion -- Apoapsis in solar orbit.

Apoapsis -- The farthest point in an orbit from the body being orbited.

Apogee -- Apoapsis in Earth orbit.

Apollo -- A class of Earth-crossing asteroid.

Apolune -- Apoapsis in lunar orbit.

Apselene -- Apoapsis in lunar orbit.

Argument -- Angular distance.

Argument of periapsis -- The argument (angular distance) of periapsis from the ascending node.

Ascending node -- The point at which an orbit crosses a reference plane (such as a planet's equatorial plane or the ecliptic plane) going north.

Asteroids -- Small bodies composed of rock and metal in orbit about the sun.

Aten -- A class of Earth-crossing asteroid.

Attometer -- 10-18 meter.

Astronomical Twilight -- For technical definition, please follow this link to the U.S. Naval Observatory Astronomical Applications website.

AU -- Astronomical Unit, based on the mean Earth-to-sun distance, 149,597,870 km. Refer to "Units of Measure" section for complete information.

AZ -- Azimuth.

B

B -- Bel, a unit of ratio equal to ten decibels. Named in honor of telecommunications pioneer Alexander Graham Bell.

B Ring -- The middle of the three rings of Saturn that are easily

SPACE TRAVEL GLOSSARY

seen in a small telescope. Barycenter -- The common center of mass about which two or more bodies revolve.

Billion -- In the U.S., 109. In other countries using SI, 1012.

Bi-phase -- A modulation scheme in which data symbols are represented by a shift from one phase to another. See Chapter 10.

BOT -- Beginning Of Track, used in DSN operations.

BPS -- Bits Per Second, same as Baud rate.

BSF -- Basics of Space Flight (this document).

BVR -- DSN Block Five (V) Receiver.

BWG -- Beam waveguide 34-m DSS, the DSN's newest DSS design.

C

c -- The speed of light, 299,792 km per second.

C-band -- A range of microwave radio frequencies in the neighborhood of 4 to 8 GHz.

C Ring -- The innermost of the three rings of Saturn that are easily seen in a small telescope.

Caltech -- The California Institute of Technology.

Carrier -- The main frequency of a radio signal generated by a transmitter prior to application of any modulation.

Cassegrain -- Reflecting scheme in antennas and telescopes having a primary and a secondary reflecting surface to "fold" the EMF back to a focus near the primary reflector.

CCD -- Charge Coupled Device, a solid-state imaging detector.

C&DH -- Command and Data Handling subsystem on board a spacecraft, similar to CDS.

CCS -- Computer Command subsystem on board a spacecraft, similar to CDS.

CCSDS -- Consultative Committee for Space Data Systems, developer of standards for spacecraft uplink and downlink, including packets.

CDR -- GCF central data recorder.

CDS -- Command and Data Subsystem onboard a spacecraft.

CDSCC -- DSN's Canberra Deep Space Communications Complex in Australia.

CDU -- Command Detector Unit onboard a spacecraft.

Centrifugal force -- The outward-tending apparent force of a body revolving around another body.

Centimeter -- 10-2 meter.

Centripetal acceleration -- The inward acceleration of a body revolving around another body.

CGPM -- General Conference of Weights and Measures, Sevres France. The abbreviation is from the French. CGPM is the source for the multiplier names (kilo, mega, giga, etc.) listed in this document.

Channel -- In telemetry, one particular measurement to which changing values may be assigned. See Chapter 10.

CIT -- California Institute of Technology, Caltech.

Civil Twilight -- For technical definition, please follow this link to the U.S. Naval Observatory Astronomical Applications website.

Clarke orbit -- Geostationary orbit.

CMC -- Complex Monitor and Control, a subsystem at DSCCs.

CMD -- DSN Command System. Also, Command data.

CNES -- Centre National d'Études Spatiales, France.

Conjunction -- A configuration in which two celestial bodies have their least apparent separation.

Coherent -- Two-way communications mode wherein the spacecraft generates its downlink frequency based upon the frequency of the uplink it receives.

Coma -- The cloud of diffuse material surrounding the nucleus of a comet.

Comets -- Small bodies composed of ice and rock in various

SPACE TRAVEL GLOSSARY

orbits about the sun.

CRAF -- Comet Rendezvous / Asteroid Flyby mission, cancelled.

CRS -- Cosmic Ray Subsystem, high-energy particle instrument on Voyager.

CRT -- Cathode ray tube video display device.

D

dB -- Decibel, an expression of ratio (usually that of power levels) in the form of log base 10. A reference may be specified, for example, dBm is referenced to milliwatts, dBW is referenced to Watts, etc. Example: 20 dBm = 1020/10 = 102 = 100 milliwatts.

DC -- Direct current.

DC -- The DSN Downlink Channel, several of which are in each DSN Downlink Tracking & Telemetry subsystem, DTT.

DCC -- The DSN Downlink Channel Controller, one of which is in each DSN Downlink Channel, DC.

DCPC -- The DSN Downlink Channel Processor Cabinet, one of which contains a DSN Downlink Channel, DC.

DEC -- Declination.

Decibel -- dB, an expression of ratio (see dB, above). One tenth of a Bel. See NIST website for further definition.

Declination -- The measure of a celestial body's apparent height above or below the celestial equator.

Density -- Mass per unit volume. For example, the density of water can be stated as 1 gram/cm3.

Descending node -- The point at which an orbit crosses a reference plane (such as a planet's equatorial plane or the ecliptic plane) going south.

DKF -- DSN keyword file, also known as KWF.

Doppler effect -- The effect on frequency imposed by relative motion between transmitter and receiver

Downlink -- Signal received from a spacecraft.

DSOT -- Data System Operations Team, part of the DSMS staff.

DSCC -- Deep Space Communications Complex, one of three DSN tracking sites at Goldstone, California; Madrid, Spain; and Canberra, Australia; spaced about equally around the Earth for continuous tracking of deep-space vehicles.

DSMS -- Deep Space Mission System, the system of computers, software, networks, and procedures that processes data from the DSN at JPL.

DSN -- Deep Space Network, NASA's worldwide spacecraft tracking facility managed and operated by JPL.

DSS -- Deep Space Station, the antenna and front-end equipment at DSCCs.

DTT -- The DSN Downlink Tracking & Telemetry subsystem.

Dyne -- A unit of force equal to the force required to accelerate a 1-g mass 1 cm per second per second. Compare with Newton.

E

E -- East.

E -- Exa, a multiplier, x10^{18} from the Greek "hex" (six, the "h" is dropped). The reference to six is because this is the sixth multiplier in the series k, M, G, T, P, E. See the entry for CGPM.

Earth -- Third planet from the sun, a terrestrial planet.

Eccentricity -- The distance between the foci of an ellipse divided by the major axis.

Ecliptic -- The plane in which Earth orbits the sun and in which solar and lunar eclipses occur.

EDL -- (Atmospheric) Entry, Descent, and Landing.

EDR -- Experiment Data Record.

EHz -- ExaHertz (10^{18} Hz)

EL -- Elevation.

Ellipse -- A closed plane curve generated in such a way that the sums of its distances from the two fixed points (the foci) is

SPACE TRAVEL GLOSSARY

constant.

ELV -- Expendable launch vehicle.

EM -- Electromagnetic.

EMF -- Electromagnetic force (radiation).

EMR -- Electromagnetic radiation.

EOT -- End Of Track, used in DSN operations.

Equator -- An imaginary circle around a body which is everywhere equidistant from the poles, defining the boundary between the northern and southern hemispheres.

ERC -- NASA's Educator Resource Centers.

ERT -- Earth-received time, UTC of an event at DSN receive-time, equal to SCET plus OWLT.

ESA -- European Space Agency.

ESP -- Extra-Solar Planet, a planet orbiting a star other than the Sun. See also Exoplanet.

ET -- Ephemeris time, a measurement of time defined by orbital motions. Equates to Mean Solar Time corrected for irregularities in Earth's motions. Obsolete, replaced by TT, Terrestrial Time.

eV -- Electron volt, a measure of the energy of subatomic particles.

Exoplanet -- Extrasolar planet. A planet orbiting a star other than the sun.

Extrasolar planet -- A planet orbiting a star other than the sun. Exoplanet.

F

f, F -- Force. Two commonly used units of force are the Newton and the dyne. Force = Mass X Acceleration.

FDS -- Flight Data Subsystem.

FE -- Far Encounter phase of mission operations.

Femtometer -- 10-15 meter.

Fluorescence -- The phenomenon of emitting light upon absorbing radiation of an invisible wavelength.

fm -- Femtometer (10-15 m)

FM -- Frequency modulation.

FTS -- DSN Frequency and Timing System. Also, frequency and timing data.

FY -- Fiscal year.

G

G -- Universal Constant of Gravitation. Its tiny value (G = 6.6726 x 10-11 Nm2/kg2) is unchanging throughout the universe.

G -- Giga, a multiplier, x109, from the Latin "gigas" (giant). See the entry for CGPM.

g -- Acceleration due to a body's gravity. Constant at any given place, the value of g varies from object to object (e.g. planets), and also with the distance from the center of the object. The relationship between the two constants is: g = GM/r2 where r is the radius of separation between the masses' centers, and M is the mass of the primary body (e.g. a planet). At Earth's surface, the value of g = 9.8 meters per second per second (9.8m/s2). See also weight.

g -- Gram, a thousandth of the metric standard unit of mass (see kg). The gram was originally based upon the weight of a cubic centimeter of water, which still approximates the current value.

Gal -- Unit of gravity field measurement corresponding to a gravitational acceleration of 1 cm/sec2.

Galaxy -- One of billions of systems, each composed of numerous stars, nebulae, and dust.

Galilean satellites -- The four large satellites of Jupiter so named because Galileo discovered them when he turned his telescope toward Jupiter: Io, Europa, Ganymede, and Callisto.

Gamma rays -- Electromagnetic radiation in the neighborhood of 100 femtometers wavelength.

SPACE TRAVEL GLOSSARY

GCF -- Ground Communications Facilities, provides data and voice communications between JPL and the three DSCCs.

GDS -- Ground Data System, encompasses DSN, GCF, DSMS, and project data processing systems.

GDSCC -- DSN's Goldstone Deep Space Communications Complex in California.

GEO -- Geosynchronous Earth Orbit.

Geostationary -- A geosynchronous equatorial circular orbit. Also called Clarke orbit.

Geosynchronous -- A direct, circular, low inclination orbit about the Earth having a period of 23 hours 56 minutes 4 seconds.

GHz -- Gigahertz (109 Hz).

GLL -- The Galileo spacecraft.

GMT -- Greenwich Mean Time. Obsolete. UT, Universal Time is preferred.

Gravitation -- The mutual attraction of all masses in the universe. Newton's Law of Universal Gravitation holds that every two bodies attract each other with a force that is directly proportional to the product of their masses, and inversely proportional to the square of the distance between them. This relation is given by the formula at right, where F is the force of attraction between the two objects, given G the Universal Constant of Gravitation, masses m1 and m2, and d distance. Also stated as $F_g = GMm/r^2$ where F_g is the force of gravitational attraction, M the larger of the two masses, m the smaller mass, and r the radius of separation of the centers of the masses. See also weight.

Gravitational waves -- Einsteinian distortions of the space-time medium predicted by general relativity theory (not yet directly detected as of March 2010). (Not to be confused with gravity waves, see below.)

Gravity assist -- Technique whereby a spacecraft takes angular momentum from a planet's solar orbit (or a satellite's orbit) to accelerate the spacecraft, or the reverse.

Gravity waves -- Certain dynamical features in a planet's atmosphere (not to be confused with gravitational waves, see above).

Great circle -- An imaginary circle on the surface of a sphere whose center is at the center of the sphere.

GSSR -- Goldstone Solar System Radar, a technique which uses very high-power X and S-band transmitters at DSS 14 to illuminate solar system objects for imaging.

GTL -- Geotail spacecraft.

GTO -- Geostationary (or geosynchronous) Transfer Orbit.

H

HA -- Hour Angle.

Halo orbit -- A spacecraft's pattern of controlled drift about an unstable Lagrange point (L1 or L2 for example) while in orbit about the primary body (e.g. the Sun).

HEF -- DSN's high-efficiency 34-m DSS, replaces STD DSSs.

Heliocentric -- Sun-centered.

Heliopause -- The boundary theorized to be roughly circular or teardrop-shaped, marking the edge of the sun's influence, perhaps 100 AU from the sun.

Heliosphere -- The space within the boundary of the heliopause, containing the sun and solar system.

HEMT -- High-electron-mobility transistor, a low-noise amplifier used in DSN.

HGA -- High-Gain Antenna onboard a spacecraft.

Hohmann Transfer Orbit -- Interplanetary trajectory using the least amount of propulsive energy. See Chapter 4.

Horizon -- The line marking the apparent junction of Earth and sky. For the technical definition, please follow this link to the U.S. Naval Observatory's Astronomical Applications.

h -- Hour, 60 minutes of time.

Hour Angle -- The angular distance of a celestial object

SPACE TRAVEL GLOSSARY

measured westward along the celestial equator from the zenith crossing. In effect, HA represents the RA for a particular location and time of day.

I

ICE -- International Cometary Explorer spacecraft.

ICRF -- International Celestial Reference Frame. The realization of the ICRS provided by the adopted positions of extragalactic objects. Link.

ICRS -- International Celestial Reference System. Conceptual basis for celestial positions, aligned with respect to extremely distant objects and utilizing the theory of general relativity. Link.

IERS -- International Earth Rotation and Reference Systems Service. Link.

IF -- Intermediate Frequency. In a radio system, a selected processing frequency between RF (Radio Frequency) and the end product (e.g. audio frequency).

Inclination -- The angular distance of the orbital plane from the plane of the planet's equator, stated in degrees.

IND -- JPL's Interplanetary Network Directorate, formerly IPN-ISD.

Inferior planet -- Planet which orbits closer to the Sun than the Earth's orbit.

Inferior conjunction -- Alignment of Earth, sun, and an inferior planet on the same side of the sun.

Ion -- A charged particle consisting of an atom stripped of one or more of its electrons.

IPAC -- Infrared Processing and Analysis Center at Caltech campus on Wilson Avenue in Pasadena.

IPC -- Information Processing Center, JPL's computing center on Woodbury Avenue in Pasadena.

IPN-ISD -- (Obsolete. See IND) JPL's Interplanetary Network and Information Systems Directorate, formerly TMOD.

IR -- Infrared, meaning "below red" radiation. Electromagnetic radiation in the neighborhood of 100 micrometers wavelength.

IRAS -- Infrared Astronomical Satellite.

ISM -- Interstellas medium.

ISO -- International Standards Organization.

ISOE -- Integrated Sequence of Events.

Isotropic -- Having uniform properties in all directions.

IUS -- Inertial Upper Stage.

J

JGR -- Journal Of Geophysical Research.

Jovian -- Jupiter-like planets, the gas giants Jupiter, Saturn, Uranus, and Neptune.

JPL -- Jet Propulsion Laboratory, operating division of the California Institute of Technology.

Jupiter -- Fifth planet from the sun, a gas giant or Jovian planet.

K

k -- Kilo, a multiplier, x103 from the Greek "khilioi" (thousand). See the entry for CGPM.

K -- Kelvin, the SI base unit of thermodynamic temperature.

K-band -- A range of microwave radio frequencies in the neighborhood of 12 to 40 GHz.

Keyhole -- An area in the sky where an antenna cannot track a spacecraft because the required angular rates would be too high. Mechanical limitations may also contribute to keyhole size. Discussed in depth under Chapter 2.

kHz -- kilohertz.

Kilogram (kg) -- the SI base unit of mass, based on the mass of a metal cylinder kept in France. See also g (gram).

Kilometer -- 103 meter.

Klystron -- A microwave travelling wave tube power amplifier

SPACE TRAVEL GLOSSARY

used in transmitters.

km -- Kilometers.

KSC -- Kennedy Space Center, Cape Canaveral, Florida.

KWF -- Keyword file of events listing DSN station activity. Also known as DKF, DSN keyword file.

Kuiper belt -- A disk-shaped region about 30 to 100 AU from the sun considered to be the source of the short-period comets.

L

Lagrange points -- Five points with respect to an orbit which a body can stably occupy. Designated L1 through L5. See Chapter 5.

LAN -- Local area network for inter-computer communications.

Large Magellanic Cloud -- LMC, the larger of two small galaxies orbiting nearby our Milky Way galaxy, which are visible from the southern hemisphere.

Laser -- Light Amplification by Stimulated Emission of Radiation. Compare with Maser.

Latitude -- Circles in parallel planes to that of the equator defining north-south measurements, also called parallels.

L-band -- A range of microwave radio frequencies in the neighborhood of 1 to 2 GHz.

LCP -- Left-hand circular polarization.

Leap Second -- A second which may be added or subtracted to adjust UTC at either, both, or neither, of two specific opportunities each year.

Leap Year -- Every fourth year, in which a 366th day is added since the Earth's revolution takes 365 days 5 hr 49 min.

LECP -- Low-Energy Charged-Particular Detector onboard a spacecraft.

LEO -- Low Equatorial Orbit.

LGA -- Low-Gain Antenna onboard a spacecraft.

Light -- Electromagnetic radiation in the neighborhood of 1

nanometer wavelength.

Light speed -- 299,792 km per second, the constant c.

Light time -- The amount of time it takes light or radio signals to travel a certain distance at light speed.

Light year -- A measure of distance, the distance light travels in one year, about 63,197 AU.

LMC -- Large Magellanic Cloud, the larger of two small galaxies orbiting nearby our Milky Way galaxy, which are visible from the southern hemisphere.

LMC -- Link Monitor and Control subsystem at the SPCs within the DSN DSCCs.

LNA -- Low-noise amplifier in DSN, either a maser or a HEMT.

Local time -- Time adjusted for location around the Earth or other planets in time zones.

Longitude -- Great circles that pass through both the north and south poles, also called meridians.

LOS -- Loss Of Signal, used in DSN operations.

LOX -- Liquid oxygen.

M

m -- Meter (U.S. spelling; elsewhere metre), the international standard of linear measurement.

m -- milli- multiplier of one one-thousandth, e.g. 1 mW = 1/1000 of a Watt, mm = 1/1000 meter.

m, M -- Mass. The kilogram is the standard unit of mass. Mass = Acceleration / Force.

M -- Mega, a multiplier, x106 (million) from the Greek "megas" (great). See the entry for CGPM.

M100 -- Messier Catalog entry number 100 is a spiral galaxy in the Virgo cluster seen face-on from our solar system.

Major axis -- The maximum diameter of an ellipse.

Mars -- Fourth planet from the sun, a terrestrial planet.

SPACE TRAVEL GLOSSARY

MC-cubed -- Mission Control and Computing Center at JPL (outdated).

MCCC -- Mission Control and Computing Center at JPL (outdated).

MCD -- DSN's maximum-likelyhood convolutional decoder, the Viterbi decoder.

MCT -- Mission Control Team, JPL Section 368 mission execution real-time operations.

MDSCC -- DSN's Madrid Deep Space Communications Complex in Spain.

Mean solar time -- Time based on an average of the variations caused by Earth's non-circular orbit. The 24-hour day is based on mean solar time.

Mercury -- First planet from the sun, a terrestrial planet.

Meridians -- Great circles that pass through both the north and south poles, also called lines of longitude.

MESUR -- The Mars Environmental Survey project at JPL, the engineering prototype of which was originally called MESUR Pathfinder, later Mars Pathfinder.

Meteor -- A meteoroid which is in the process of entering Earth's atmosphere. It is called a meteorite after landing.

Meteorite -- Rocky or metallic material which has fallen to Earth or to another planet.

Meteoroid -- Small bodies in orbit about the sun which are candidates for falling to Earth or to another planet.

MGA -- Medium-Gain Antenna onboard a spacecraft.

MGN -- The Magellan spacecraft.

MGSO -- (Obsolete. See TMOD) JPL's Multimission Ground Systems Office.

MHz -- Megahertz (106 Hz).

Micrometer -- μm, 10-6 meter.

Micron -- Obsolete terms for micrometer, μm (10-6 m).

Milky Way -- The galaxy which includes the sun and Earth.

Millimeter -- 10-3 meter.

MIT -- Massachusetts Institute of Technology.

MLI -- Multi-layer insulation (spacecraft blanketing). See Chapter 11.

mm -- millimeter (10-3 m).

MO -- The Mars Observer spacecraft.

Modulation -- The process of modifying a radio frequency by shifting its phase, frequency, or amplitude to carry information.

MON -- DSN Monitor System. Also, monitor data.

Moon -- A small natural body which orbits a larger one. A natural satellite. Capitalized, the Earth's natural satellite.

Moonrise -- For technical definition, please follow this link to the U.S. Naval Observatory Astronomical Applications website.

Moonset -- For technical definition, please follow this link to the U.S. Naval Observatory Astronomical Applications website.

MOSO -- Multimission Operations Systems Office at JPL.

MR -- Mars relay.

μm -- Micrometer (10-6 m).

Multiplexing -- A scheme for delivering many different measurements in one data stream.

N

N -- Newton, the SI unit of force equal to that required to accelerate a 1-kg mass 1 m per second per second (1m/sec2). Compare with dyne.

N -- North.

Nadir -- The direction from a spacecraft directly down toward the center of a planet. Opposite the zenith.

NASA -- National Aeronautics and Space Administration.

Nautical Twilight -- For technical definition, please follow this

SPACE TRAVEL GLOSSARY

link to the U.S. Naval Observatory Astronomical Applications website.

NE -- Near Encounter phase in flyby mission operations.

Neptune -- Eighth planet from the sun, a gas giant or Jovian planet.

NiCad -- Nickel-cadmium rechargable battery.

NIMS -- Near-Infrared Mapping Spectrometer onboard the Galileo spacecraft.

NIST -- National Institute of Standards.

nm -- Nanometer (10-9 m).

nm -- Nautical Mile, equal to the distance spanned by one minute of arc in latitude, 1.852 km.

NMC -- Network Monitor and Control subsystem in DSN.

NOCC -- DSN Network Operations Control Center at JPL.

Nodes -- Points where an orbit crosses a reference plane.

Non-coherent -- Communications mode wherein a spacecraft generates its downlink frequency independent of any uplink frequency.

Nucleus -- The central body of a comet.

Nutation -- A small nodding motion in a rotating body. Earth's nutation has a period of 18.6 years and an amplitude of 9.2 arc seconds.

NRZ -- Non-return to zero. Modulation scheme in which a phase deviation is held for a period of time in order to represent a data symbol. See Chapter 10.

NSP -- DSN Network Simplification Project. A project that re-engineered the DSN to consolidate seven data systems into two data systems that handle the same data types.

O

OB -- Observatory phase in flyby mission operations encounter period.

One-way -- Communications mode consisting only of downlink

received from a spacecraft.

Oort cloud -- A large number of comets theorized to orbit the sun in the neighborhood of 50,000 AU.

OPCT -- Operations Planning and Control Team at JPL, "OPSCON." Obsolete, replaced by DSOT, Data Systems Operations Team.

Opposition -- Configuration in which one celestial body is opposite another in the sky. A planet is in opposition when it is 180 degrees away from the sun as viewed from another planet (such as Earth). For example, Saturn is at opposition when it is directly overhead at midnight on Earth.

OPNAV -- Optical Navigation (images).

OSI -- ISO's Open Systems Interconnection protocol suite.

OSR -- Optical Solar Reflector, thermal control component onboard a spacecraft.

OSS -- Office Of Space Science, NASA. Obsolete, replaced by Science Mission Directorate (SMD).

OSSA -- Office Of Space Science and Applications, NASA (Obsolete, see OSS).

OTM -- Orbit Trim Maneuver, spacecraft propulsive maneuver.

OWLT -- One-Way Light Time, elapsed time between Earth and spacecraft or solar system body.

P

P -- Peta, a multiplier, x1015, from the Greek "pente" (five, the "n" is dropped). The reference to five is because this is the fifth multiplier in the series k, M, G, T, P. See the entry for CGPM.

Packet -- A quantity of data used as the basis for multiplexing, for example in accordance with CCSDS.

PAM -- Payload Assist Module upper stage.

Parallels -- Circles in parallel planes to that of the equator defining north-south measurements, also called lines of latitude.

Pathfinder -- The Mars Environmental Survey (MESUR)

SPACE TRAVEL GLOSSARY

engineering prototype later named Mars Pathfinder.

PDS -- Planetary Data System.

PDT -- Pacific Daylight Time.

PE -- Post Encounter phase in flyby mission operations.

Periapsis -- The point in an orbit closest to the body being orbited.

Perigee -- Periapsis in Earth orbit.

Perichron -- Periapsis in Saturn orbit.

Perihelion -- Periapsis in solar orbit.

Perijove -- Periapsis in Jupiter orbit.

Perilune -- Periapsis in lunar orbit.

Periselene -- Periapsis in lunar orbit.

Phase -- The angular distance between peaks or troughs of two waveforms of similar frequency.

Phase -- The particular appearance of a body's state of illumination, such as the full or crescent phases of the Moon.

Phase -- Any one of several predefined periods in a mission or other activity.

Photovoltaic -- Materials that convert light into electric current.

PHz -- Petahertz (10^{15} Hz).

PI -- Principal Investigator, scientist in charge of an experiment.

Picometer -- 10^{-12} meter.

PIO -- JPL's Public Information Office.

Plasma -- Electrically conductive fourth state of matter (other than solid, liquid, or gas), consisting of ions and electrons.

PLL -- Phase-lock-loop circuitry in telecommunications technology.

Pluto -- Ninth planet from the sun, sometimes classified as a small terrestrial planet.

pm -- Picometer (10^{-12} m).

PM -- Post meridiem (Latin: after midday), afternoon.

PN10 -- Pioneer 10 spacecraft.

PN11 -- Pioneer 11 spacecraft.

Prograde -- Orbit in which the spacecraft moves in the same direction as the planet rotates. See retrograde.

PST -- Pacific Standard Time.

PSU -- Pyrotechnic Switching Unit onboard a spacecraft.

Q

Quasar -- Quasi-stellar object observed mainly in radio waves. Quasars are extragalactic objects believed to be the very distant centers of active galaxies.

R

RA -- Right Ascension.

Radian -- Unit of angular measurement equal to the angle at the center of a circle subtended by an arc equal in length to the radius. Equals about 57.296 degrees.

RAM -- Random Access Memory.

RCP -- Right-hand circular polarization.

Red dwarf -- A small star, on the order of 100 times the mass of Jupiter.

Reflection -- The deflection or bouncing of electromagnetic waves when they encounter a surface.

Refraction -- The deflection or bending of electromagnetic waves when they pass from one kind of transparent medium into another.

REM -- Receiver Equipment Monitor within the Downlink Channel (DC) of the Downlink Tracking & Telemetry subsystem (DTT).

Retrograde -- Orbit in which the spacecraft moves in the opposite direction from the planet's rotatation. See prograde.

RF -- Radio Frequency.

SPACE TRAVEL GLOSSARY

RFI -- Radio Frequency Interference.

Right Ascension -- The angular distance of a celestial object measured in hours, minutes, and seconds along the celestial equator eastward from the vernal equinox.

Rise -- As in ascending above the horizon, for the technical definition, please follow this link to the U.S. Naval Observatory's Astronomical Applications.

RNS -- GCF reliable network service.

ROM -- Read Only Memory.

RPIF -- Regional Planetary Imaging Data Facilities.

RRP -- DSN Receiver & Ranging Processor within the Downlink Channel (DC) of the Downlink Tracking & Telemetry subsystem (DTT).

RS -- DSN Radio Science System. Also, radio science data.

RTG -- Radioisotope Thermo-Electric Generator onboard a spacecraft.

RTLT -- Round-Trip Light Time, elapsed time roughly equal to 2 x OWLT.

S

S -- South.

s -- Second, the SI base unit of time (see this extensive definition).

SA -- Solar Array, photovoltaic panels onboard a spacecraft.

SAF -- Spacecraft Assembly Facility, JPL Building 179.

SAR -- Synthetic Aperture Radar

Satellite -- A small body which orbits a larger one. A natural or an artificial moon. Earth-orbiting spacecraft are called satellites. While deep-space vehicles are technically satellites of the sun or of another planet, or of the galactic center, they are generally called spacecraft instead of satellites.

Saturn -- Sixth planet from the sun, a gas giant or Jovian planet.

S-band -- A range of microwave radio frequencies in the neighborhood of 2 to 4 GHz.

SC -- Steering Committee.

SCET -- Spacecraft Event Time, equal to ERT minus OWLT.

SCLK -- Spacecraft Clock Time, a counter onboard a spacecraft.

Sec -- Abbreviation for Second.

Second -- the SI base unit of time. See this extensive definition.

SEDR -- Supplementary Experiment Data Record.

SEF -- Spacecraft event file.

SEGS -- Sequence of Events Generation Subsystem.

Semi-major axis -- Half the distance of an ellipse's maximum diameter, the distance from the center of the ellipse to one end.

Set -- As in going below the horizon, for the technical definition, please follow this link to the U.S. Naval Observatory's Astronomical Applications.

SFOF -- Space Flight Operations Facility, Buildings 230 and 264 at JPL.

SFOS -- Space Flight Operations Schedule, product of SEGS.

Shepherd moons -- Moons which gravitationally confine ring particles.

SI -- The International System of Units (metric system). See also Units of Measure.

SI base unit -- One of seven SI units of measure from which all the other SI units are derived. See SI derived unit. See also Units of Measure.

SI derived unit -- One of many SI units of measure expressed as relationships of the SI base units. For example, the watt, W, is the SI derived unit of power. It is equal to joules per second. $W = m^2 \: kg \: s^{-3}$ (Note: the joule, J, is the SI derived unit for energy, work, or quantity of heat.) See also Units of Measure.

Sidereal time -- Time relative to the stars other than the sun.

SPACE TRAVEL GLOSSARY

SIRTF -- Space Infrared Telescope Facility.

SMC -- Small Magellanic Cloud, the smaller of two small galaxies orbiting nearby our Milky Way galaxy, which are visible from the southern hemisphere.

SMD -- Science Mission Directorate, NASA (previously Office Of Space Science, OSS).

SOE -- Sequence of Events.

Solar wind -- Flow of lightweight ions and electrons (which together comprise plasma) thrown from the sun.

SNR -- Signal-to-Noise Ratio.

SPC -- Signal Processing Center at each DSCC.

Specific Impulse -- A measurement of a rocket's relative performance. Expressed in seconds, the number of which a rocket can produce one pound of thrust from one pound of fuel. The higher the specific impulse, the less fuel required to produce a given amount of thrust.

Spectrum -- A range of frequencies or wavelengths.

SSA -- Solid State Amplifier in a spacecraft telecommunications subsystem, the final stage of amplification for downlink.

SSI -- Solid State Imaging Subsystem, the CCD-based cameras on Galileo.

SSI -- Space Services, Inc., Houston, manufacturers of the Conestoga launch vehicle.

STD -- Standard 34-m DSS, retired from DSN service.

STS -- Space Transportation System (Space Shuttle).

Subcarrier -- Modulation applied to a carrier which is itself modulated with information-carrying variations.

Sunrise -- For technical definition, please follow this link to the U.S. Naval Observatory Astronomical Applications website.

Sunset -- For technical definition, please follow this link to the U.S. Naval Observatory Astronomical Applications website.

Sun synchronous orbit -- A spacecraft orbit that precesses,

wherein the location of periapsis changes with respect to the planet's surface so as to keep the periapsis location near the same local time on the planet each orbit. See walking orbit.

Superior planet -- Planet which orbits farther from the sun than Earth's orbit.

Superior conjunction -- Alignment between Earth and a planet on the far side of the sun.

SWG -- Science Working Group.

T

TAU -- Thousand AU Mission.

TCM -- Trajectory Correction Maneuver, spacecraft propulsive maneuver.

TDM -- Time-division multiplexing.

Termination shock -- Shock at which the solar wind is thought to slow to subsonic speed, well inside the heleopause.

T -- Tera, a multiplier x1012, from the Greek teras (monster). See the entry for CGPM.

Terrestrial planet -- One of the four inner Earth-like planets.

Three-way -- Coherent communications mode wherein a DSS receives a downlink whose frequency is based upon the frequency of an uplink provided by another DSS.

TMOD -- (Obsolete. See IPN-ISD) JPL's Telecommunications and Mission Operations Directorate. Formerly MGSO.

THz -- Terahertz (1012 Hz).

TLM -- DSN Telemetry data.

TLP -- DSN Telemetry Processor within the DTT Downlink Channel.

TOS -- Transfer Orbit Stage, upper stage.

Transducer -- Device for changing one kind of energy into another, typically from heat, position, or pressure into a varying electrical voltage or vice-versa, such as a microphone or speaker.

SPACE TRAVEL GLOSSARY

Transit -- For technical definition, please follow this link to the U.S. Naval Observatory Astronomical Applications website.

Transponder -- Electronic device which combines a transmitter and a receiver.

TRC -- NASA's Teacher Resource Centers. Obsolete, now called Educator Resource Centers, ERC.

TRK -- DSN Tracking System. Also, Tracking data.

TRM -- Transmission Time, UTC Earth time of uplink.

True anomaly -- The angular distance of a point in an orbit past the point of periapsis, measured in degrees.

Twilight -- For technical definition, please follow this link to the U.S. Naval Observatory Astronomical Applications website.

TWNC -- Two-Way Non-Coherent mode, in which a spacecraft's downlink is not based upon a received uplink from DSN.

Two-way -- Communications mode consisting of downlink received from a spacecraft while uplink is being received at the spacecraft. See also coherent.

TWT -- Traveling Wave Tube, downlink power amplifier in a spacecraft telecommunications subsystem, the final stage of amplification for downlink (same unit as TWTA).

TWTA -- Traveling Wave Tube Amplifier, downlink power amplifier in a spacecraft telecommunications subsystem, the final stage of amplification for downlink (same unit as TWT).

TXR -- DSN's DSCC Transmitter assembly.

U

UHF -- Ultra-high frequency (around 300MHz).

μm -- Micrometer (10-6 m).

ULS -- Ulysses spacecraft.

Uplink -- Signal sent to a spacecraft.

UPL -- The DSN Uplink Tracking & Command subsystem.

Uranus -- Seventh planet from the sun, a gas giant or Jovian planet.

USO -- Ultra Stable Oscillator, in a spacecraft telecommunications subsystem.

UT -- Universal Time, also called Zulu (Z) time, previously Greenwich Mean Time. UT is based on the imaginary "mean sun," which averages out the effects on the length of the solar day caused by Earth's slightly non-circular orbit about the sun. UT is not updated with leap seconds as is UTC.

UTC -- Coordinated Universal Time, the world-wide scientific standard of timekeeping. It is based upon carefully maintained atomic clocks and is highly stable. Its rate does not change by more than about 100 picoseconds per day. The addition or subtraction of leap seconds, as necessary, at two opportunities every year adjusts UTC for irregularities in Earth's rotation. The U.S. Naval Observatory website provides information in depth on the derivation of UTC.

UV -- Ultraviolet (meaning "above violet") radiation. Electromagnetic radiation in the neighborhood of 100 nanometers wavelength.

UWV -- DSN Microwave subsystem in DSSs which includes waveguides, waveguide switches, LNAs, polarization filters, etc.

V

Velocity -- A vector quantity whose magnitude is a body's speed and whose direction is the body's direction of motion.

Venus -- Second planet from the sun, a terrestrial planet.

VGR1 -- Voyager 1 spacecraft.

VGR2 -- Voyager 2 spacecraft.

VLBI -- DSN Very Long Baseline Interferometry System. Also, VLBI data. Link.

W

W -- Watt, a measure of electrical power equal to potential in volts times current in amps.

W -- West.

Walking orbit -- A spacecraft orbit that precesses, wherein

the location of periapsis changes with respect to the planet's surface in a useful way. See sun-synchronous.

Wavelength -- The distance that a wave from a single oscillation of electromagnetic radiation will propagate during the time required for one oscillation.

Weight -- The gravitational force exerted on an object of a certain mass. The weight of mass m is mg Newtons, where g is the local acceleration due to a body's gravity.

WWW -- World-Wide Web.

Z

X-band -- A range of microwave radio frequencies in the neighborhood of 8 to 12 GHz.

X-ray -- Electromagnetic radiation in the neighborhood of 100 picometer wavelength.

Y

Y -- Yotta, a multiplier, x1024 from the second-to-last letter of the Latin alphabet. See the entry for CGPM.

Z

Z -- Zetta, a multiplier, x1021 from the last letter of the Latin alphabet. See the entry for CGPM.

Z -- Zulu in phonetic alphabet, stands for UT, Universal Time.

Zenith -- The point on the celestial sphere directly above the observer. Opposite the nadir. Crater: A round impression left in a planet or satellite from a meteoroid.

THE MOON LANDING HOAX